Virology

Virology

Khushboo Chaudhary

Delve
PUBLISHING

www.delvepublishing.com

Virology

Khushboo Chaudhary

Delve Publishing
224 Shoreacres Road
Burlington, ON L7L 2H2
Canada
www.delvepublishing.com

Email: orders@arclereducation.com

ISBN: 978-1-77469-069-7 (Hardcover)

ABOUT THE AUTHOR

Dr. Khushboo Chaudhary is presently working as a Technical officer-I in Translational Health Science and Technology Institute, Faridabad Haryana, India and having one year of teaching experience and eight years of research experience. Previously, she worked on the PPR virus and New Castle Disease virus and isolates many other viruses from animal outbreak samples. She has published several research papers in international and national journals. She has published five international textbooks. She has got seven best paper and poster presentation awards from the Indian Society of Genetics and Biotechnology Research and Development and received the president appreciation awards also in the International conference. She has got the best poster award by ISSGPU Central Institute Research on Goats, Makhdoom. She has published several gene banks in NCBI Pubmed. She has also published a research article in a virology journal. She is likely to be a co-author in several of the publications and coauthor in J. Virological Methods.

TABLE OF CONTENTS

GLOSSARY

Agarose Gel Electrophoresis: It is a process in which a matrix composed of a highly purified form of agar is used to separate large DNA and RNA molecules.

Amplification: There are formations of many copies of the DNA segment of PCR.

Annealing: The process of healing and slowly cooling of double-stranded is DNA to allow the formation of hybrid DNA or DNA-RNA molecules.

Antibiotic: A biological substance produced by one organism inhibiting the growth of another organism.

Antibody: A protein synthesized by B lymphocytes that recognizes a specific site on an antigen.

Antigen: A compound that induces the production of antibodies.

Antiparallel orientation: Arrangement of two strands of DNA molecule oriented in opposite directions so that the 5′ PO4 end of one strand is aligned with the 3′OH end of the complementary strand.

Antisense DNA: The sequence of chromosomal DNA that is transcribed.

Antisense RNA: An RNA sequence complementary to all or part of a functional RNA.

Antisense Therapy: *In vitro* treatment of genetic disease by blocking translation of a protein with DNA or RNA sequences that is complementary to a specific mRNA.

Autoclave: An instrument used for sterilization of glassware and culture media.

cDNA Clone: A double-stranded DNA complement of mRNA Synthesised *in vitro* by using reverse transcriptase and DNA polymerase.

cDNA library: A collection of cDNA clones generated *in vitro* from the mRNA sequences of a single tissue or cell population.

Cell Culture: The cell cultures of growing cells are *in vitro* in the liquid medium.

Cell Line: The cell line is derived from a primary culture and selected at the time of the first subculture or derived from a cell population with particular attributes.

Chimera: Recombinant DNA molecules containing sequences from different organisms.

Clone: A collection of genetically identical cells or organisms derived from a common ancestor where all members have a similar genetic composition.

Consensus sequences: The nucleotide sequences that are present in the majority of genetic signals that perform a specific function.

Cosmid: A plasmid vector that consists of the COS site of phage lambda (λ) and one or more selectable markers such as an antibiotic resistance gene.

Denaturation: The separation of DNA double-stranded molecule into single strands.

Differentiation: The process of biochemical and structural changes by which cells become specialized in form and function.

DNA Construct: A suitable DNA which has been prepared for cloning purpose.

DNA polymerase: An enzyme that catalyzes the phosphodiester bond in the formation of DNA.

DNA Probe: A radiolabelled (32P) DNA segment used to hybridize the base between the probe and complementary base sequences in a DNA sample.

Electrophoresis: The method of separation of a molecule such as DNA, RNA or protein based on their relative migration applying a strong electric field.

Embryo Transfer: Implantation of embryos from donor animals or generated by *in vitro* fertilization into the uterus of the recipient animals.

Excision: Enzymatic removal of a DNA segment from a chromosome.

Exonucleases: An enzyme that digests DNA or RNA from the ends of the strand; 5′ exonuclease requires a free 5′ end and degrades the molecule in the 5′-3′ direction; 3′ exonuclease requires a free 3′ end and degrades the DNA IN 3′-5′ direction.

Fusion Product: A single protein molecule encoded by parts of two or more genes combined together as one unit.

Gene: A hereditary unit or the segment of DNA coding a specific protein.

Genetic Fingerprinting: A technique of analyzing the DNA of an individual to reveal the pattern of repetition of particular nucleotide sequences through the genome.

Hybridoma: A unique fused cells or protoplasts into a matrix.

***In vitro*:** Any process carried out in sterile cultures or measurement of biological processes outside the intact organism such as enzyme reaction.

In vivo: The natural condition in which organisms live.

Leader Sequences: It is the non-translated sequence at the 5′ end of mRNA that precedes the initiation codon.

Lysogeny: The phenomenon in which prophage survives within a host bacterium as a part of the host genome or an extrachromosomal element and does not cause lysis.

Monoclonal antibodies: It is derived from single clone cell ,which recognizes only one kind of antigen.

Oncogene: A gene that causes cancer.

Operon: A cluster of genes that are coordinately regulated.

Palindromic sequences: The complementary DNA sequences that are the same when each strand is read in the same direction e.g (5′-3). These sequences act as recognition sites for type II restriction endonuclease.

Patent: A legal document issued by a government that permits the holder the exclusive right to manufacture, use or sellng an invention for a defined period, which varies from country to country.

Plasmid: Extrachromosomal, self-replicating circular double-stranded DNA molecules containing some non-essential genes.

Primer: A short oligonucleotide that hybridizes the template strand and gives a 3′-OH end for the initiation of nucleic acid synthesis.

Recombinant: An individual whose genes on chromosomes result from one or more cross-over events.

Splicing: The gene manipulation where the one DNA molecule is attached to one another.

Transfection: DNA transfer into a eukaryotic cell.

Transposon: A DNA segment that can Jump from the locus and join a new locus on the DNA molecule by involving the enzyme transposase.

Vectors: The vector used for transferring DNA from one cell to another or DNA molecule that can carry inserted DNA and be perpetuated in a host cell.

Western Blot: A technique in which protein is a transfer from electrophoresis gen to a cellulose or nylon support membrane following electrophoresis.

Yeast Artificial Chromosomes: A vector is used for hundreds to kilobases long used for cloning of DNA fragments.

Zooblot: The cloned DNA from one species to DNA from the other organism to determine the extent to which the cloned DNA is evolutionary conserved.

ACKNOWLEDGMENT

First and foremost, we are grateful to the almighty God. We could never have done this without his blessings. We would also like to acknowledge the major we are sincerely thankful to Dr. Naveen Kumar (Principal Scientist-NRCE Hisar, Haryana, India) and Dr. Guruprasad Medigeshi (Professor), THSTI, Faridabad, Haryana for providing us research facilities and the Bioinformatics Centre, for extensive use of computational facilities. We are deeply indebted to many people who gave their valuable support in the writing of this book; to all those who provided encouragement, support, discussed things over, offered comments assisted in the editing. Above all, we want to thank our family, friends and lab mates for their passionate encouragement and support, throughout the writing of this book. We are also thankful to those entire people who directly and indirectly support the writing journey of this book. Otherwise, a huge textbook could not be prepared.

Khushboo Chaudhary

PREFACE

The book **"Virology"** has been written to serve as a textbook for undergraduate, postgraduate and research scholar of Indian and other country universities. There are many books on the subject of Virology but updated information is not incorporated in the edition. This book has covered a range of syllabus and the student is an easy way to go through this book content. Some books are available on the subject of virology but still not up to date.

While this book has prepared for students who have acquired or gain fair knowledge of **Virology** and easy presentation of language has been written for a proper understanding of each and every topic. The chapters have been so arranged to give a sequential knowledge of the introduction of viruses, virus-cell interaction, Peste des petits ruminants (PPR virus), SARS coronavirus 2, Severe acute respiratory syndrome coronavirus 2, CRISPR/Cas technology use in therapeutics applications, virology laboratory requirements and safety rules and regulations, etc, in this book with practical approaches. The examples to illustrate different points up to date have been chosen from different universities and institutions of the country.

This book would serve not only for undergraduate but postgraduate students of biotechnology and also research scholar of various universities. The numbers of scientists and authors have been quoted or fully acknowledge and referenced at the end of the book. In the preparing for this book, I have been greatly helped by several books and recent publications, which I express my acknowledgment in the references and if I have forgotten any references in my notice, please ignore. Comments and suggestions are most welcome from students as well as from many teachers' friends for corrections and improvement of this book will be greatly acknowledged. I am particularly thankful to **Dr. Naveen Kumar (Principal Scientist, Hisar)** has given me constant inspiration. Special thanks to my son **Lavyansh and** my parents **Rakesh Chaudhary and Ajesh Chaudhary who** have given full support to complete this book. Last but not least I would be thankful to readers for pointing out errors and most welcome for suggestions in the next book.

Khushboo Chaudhary

PART-I

INTRODUCTION OF VIRUSES

CHAPTER 1

VIRUS

Viruses is small intracellular obligate parasites, which contain either a DNA or RNA genome surrounded by a protective and virus-coded protein coat. Viruses may be viewed as mobile genetic elements, most probably of cellular origin and characterized by a long co-evolution of virus and host. For propagation, viruses depend on specialized host cells supplying the complex metabolic and biosynthetic machinery of eukaryotic or prokaryotic cells. A complete virus particle is called a virion. The main function of the virion is to deliver its DNA or RNA genome into the host cell so that the genome can be expressed (transcribed and translated) by the host cell. The viral genome, often with associated basic proteins, is packaged inside a symmetric protein capsid. The nucleic acid-associated protein, called nucleoprotein, together with the genome, forms the nucleocapsid. In enveloped viruses, the nucleocapsid is surrounded by a lipid bilayer derived from the modified host cell membrane and studded with an outer layer of virus envelope glycoproteins.

CLASSIFICATION OF VIRUSES

Morphology

Viruses are grouped on the basis of size and shape, chemical composition and structure of the genome, and mode of replication. Helical morphology is seen in nucleocapsids of many filamentous and pleomorphic viruses. Helical nucleocapsids consist of a helical array of capsid proteins (protomers) wrapped around a helical filament of nucleic acid. Icosahedral morphology is characteristic of the nucleocapsids of many "spherical" viruses. The number and arrangement of the capsomeres (morphologic subunits of the icosahedron) are useful in identification and classification. Many viruses also have an outer envelope.

Chemical Composition and Mode of Replication

The genome of a virus may consist of DNA or RNA, which may be single-stranded (ss) or double-stranded (ds), linear or circular. The entire genome may occupy either one nucleic acid molecule (monopartite genome) or several nucleic acid segments (multipartite genome). The different types of genome necessitate different replication strategies.

Nomenclature

Aside from physical data, genome structure and mode of replication are criteria applied in the classification and nomenclature of viruses, including the chemical composition and configuration of the nucleic acid, whether the genome is monopartite or multipartite. The genomic RNA strand of single-stranded RNA viruses is called sense (positive sense, plus sense) in orientation if it can serve as mRNA, and antisense (negative sense, minus sense) if a complementary strand is synthesized by viral RNA transcriptase to serve as mRNA. Also considered in viral classification is the site of capsid assembly and, in enveloped viruses, the site of envelopment.

Viruses are inert outside the host cell. Small viruses, e.g., polio and tobacco mosaic virus, can even be crystallized. Viruses are unable to generate energy. As obligate intracellular parasites, during replication, they fully depend on the complicated biochemical machinery of eukaryotic or prokaryotic cells. The main purpose of a virus is to deliver its genome into the host cell to allow its expression (transcription and translation) by the host cell.

A fully assembled infectious virus is called a virion. The simplest virions consist of two basic components: nucleic acid (single- or double-stranded RNA or DNA) and a protein coat, the capsid, which functions as a shell to protect the viral genome from nucleases and which during infection attaches the virion to specific receptors exposed on the prospective host cell. Capsid proteins are coded for by the virus genome. Because of its limited size the genome codes for only a few structural proteins (besides non-structural regulatory proteins involved in virus replication). Capsids are formed as single or double protein shells and consist of only one or a few structural protein species. Therefore, multiple protein copies must self assemble to form the continuous three-dimensional capsid structure. Self-assembly of virus capsids follows two basic patterns: helical symmetry, in which the protein subunits and the nucleic acid are arranged in a helix, and icosahedral symmetry, in which the protein subunits assemble into a symmetric shell that covers the nucleic acid-containing core.

Some virus families have an additional covering, called the envelope, which is usually derived in part from modified host cell membranes. Viral envelopes consist of a lipid bilayer that closely surrounds a shell of virus-encoded membrane-associated proteins. The exterior of the bilayer is studded with virus-coded, glycosylated (trans-) membrane proteins. Therefore, enveloped viruses often exhibit a fringe of glycoprotein spikes or knobs, also called peplomers. In viruses that acquire their envelope by budding through the plasma or another intracellular cell membrane, the lipid composition of the viral envelope closely reflects that of the particular host membrane. The outer capsid and the envelope proteins of viruses are glycosylated and important in determining the host range and antigenic composition of the virion. In addition to virus-specified envelope proteins, budding viruses carry also certain host cell proteins as integral constituents of the viral envelope. Virus envelopes can be considered an additional protective coat. Larger viruses often have a complex architecture consisting of both helical and isometric symmetries confined to different structural components. Small viruses, e.g., hepatitis B virus or the members of the picornavirus or parvovirus family, are orders of magnitude more resistant than are the larger complex viruses, e.g. members of the herpes or retrovirus families.

Chemical Composition and Mode of Replication

RNA Virus Genomes

RNA viruses, comprising 70% of all viruses, vary remarkably in genome structure. Because of the error rate of the enzymes involved in RNA replication, these viruses usually show much higher mutation rates than do the DNA viruses. Mutation rates of 10-4 lead to the continuous generation of virus variants which show great adaptability to new hosts. The viral RNA may be single-stranded (ss) or double-stranded (ds), and the genome may occupy a single RNA segment or be distributed on two or more separate segments (segmented genomes). In addition, the RNA strand of a single-stranded genome may be either a sense strand (+ strand), which can function as messenger RNA (mRNA) or an antisense strand (- strand), which is complementary to the sense strand and cannot function as mRNA protein translation. Sense viral RNA alone can replicate if injected into cells since it can function as mRNA and initiate translation of virus-encoded proteins. Antisense RNA, on the other hand, has no translational function and cannot per se produce viral components.

DsRNA viruses, e.g., members of the reovirus family, contain 10, 11 or 12 separate genome segments coding for 3 enzymes involved in RNA replication, 3 major capsid proteins and a number of smaller structural proteins. Each segment consists of a complementary sense and antisense strand that is hydrogen-bonded into a linear ds molecule. The replication of these viruses is complex; only the sense RNA strands are released from the infecting virion to initiate replication.

The retrovirus genome comprises two identical, plus-sense ssRNA molecules, each monomer 7-11 kb in size, that are noncovalently linked over a short terminal region. Retroviruses contain 2 envelope proteins encoded by the env-gene, 4 - 6 nonglycosylated core proteins and 3 non-structural functional proteins (reverse transcriptase, integrase, protease: RT, IN, PR, respectively) coded by the gag-gene. The RT transcribes the viral ssRNA into double- stranded, circular proviral DNA. This DNA, mediated by the viral integrase, becomes covalently bonded into the DNA of the host cell to make possible the subsequent transcription of the sense strands that eventually give rise to retrovirus progeny. After assembly and budding, retroviruses show structural and functional maturation. In immature virions, the structural proteins of the core are present as a large precursor protein shell. After proteolytic processing by the viral protease, the proteins of the mature virion are rearranged and form the dense isometric or cone-shaped core typical of the mature virion, and the particle becomes infectious.

DNA Virus Genomes

Most DNA viruses contain a single genome of linear dsDNA. The papovaviruses, comprising the polyoma-and papillomaviruses, however, have circular DNA genomes, about 5.1 and 7.8 kb pairs in size. DsDNA serves as a template both for mRNA and for self-transcription. Three or 2 structural proteins make up the papovavirus capsid: in addition, 5-6 nonstructural proteins are encoded that are functional in virus transcription, DNA replication and cell transformation. Single-stranded linear DNA, 4-6 kb in size, is found with the members of the Parvovirus family that comprises the parvo-, the erythro- and the dependoviruses. The virion contains 2-4 structural protein species which are differently derived from the same gene product. The adeno-associated virus (AAV, a dependovirus) is incapable of producing progeny virions except in the presence of helper viruses (adenovirus or herpesvirus).

It is therefore said to be replication-defective. Circular single-stranded DNA of only 1.7 to 2.3 kb is found in members of the Circovirus family, which comprise the smallest autonomously propagated viruses. The isometric capsid measures 17 nm and is composed of 2 protein species only.

Virus Classification

On the basis of shared properties viruses are grouped at different hierarchical levels of order, family, subfamily, genus and species. More than 30,000 different virus isolates are known today and grouped in more than 3,600 species, in 164 genera and 71 families. Viral morphology provides the basis for grouping viruses into families. A virus family may consist of members that replicate only in vertebrates, only in invertebrates, only in plants, or only in bacteria. Certain families contain viruses that replicate in more than one of these hosts. This section concerns only 21 families and genera of medical importance.

Besides physical properties, several factors pertaining to the mode of replication play a role in classification: the configuration of the nucleic acid (ss or ds, linear or circular), whether the genome consists of one molecule of nucleic acid or is segmented, and whether the strand of ss RNA is sense or antisense. Also considered in classification is the site of viral capsid assembly and, in enveloped viruses, the site of nucleocapsid envelopment. The major chemical and morphologic properties of the families of viruses are cause disease in humans. The use of Latinized names ending in -viridae for virus families and ending in -virus for viral genera has gained wide acceptance.

The names of subfamilies end in -virinae. Vernacular names continue to be used to describe the viruses within a genus. In this text, Latinized endings for families and subfamilies usually are not used. In the early days of virology, viruses were named according to common pathogenic properties, e.g. organ tropism and/or modes of transmission, and often also after their discoverers. From the early 1950s until the mid-1960s, when many new viruses were being discovered, it was popular to compose virus names by using sigla. Thus, the name Picornaviridae is derived from pico (small) and RNA; the name Reoviridae is derived from respiratory, enteric, and orphan viruses because the agents were found in both respiratory and enteric specimens and were not related to other classified viruses; Papovaviridae is from papilloma, polyoma, and vacuolating agent (simian virus 40 [SV40]); retrovirus is from reverse transcriptase; Hepadnaviridae is from the replication of the virus in hepatocytes and their DNA genomes, as seen in hepatitis B virus.

Hepatitis A virus is classified now in the family Picornaviridae, genus Hepatovirus. Although the current rules for nomenclature do not prohibit the introduction of new sigla, they require that the siglum be meaningful to workers in the field and be recognized by international study groups.

The names of the other families that contain viruses pathogenic for humans are derived as follows: Adenoviridae (adeno, "gland"; refers to the adenoid tissue from which the viruses were first isolated); Astroviridae (astron means star); Arenaviridae (arena "sand") describes the sandy appearance of the virion. Bunyaviridae (from Bunyamwera, the place in Africa where the type strain was isolated); Calicivirus (calix, "cup" or "goblet" from the cup-shaped depressions on the viral surfaces); Coronaviridae (corona, "crown") describes the appearance of the peplomers protruding from the viral surface; Filoviridae (from the Latin filum, "thread" or "filament") describes the morphology of these viruses. Herpesviridae (herpes, "creeping") describes the nature of the lesions; Orthomyxoviridae (ortho, "true," plus myxo "mucus," a substance for which the viruses have an affinity; Paramyxoviridae derived from para, "closely resembling" and myxo; Parvoviridae (parvus means, "small"); Poxviridae (pock means, "pustule"); Rhabdoviridae (rhabdo, "rod" describes the shape of the viruses and Togaviridae (toga, "cloak") refers to the tight viral envelope.

Major Cause of Disease

viruses are also important as a new source of therapy.

The new viruses are emerging.

Important Members of the Aquatic World

It moves organic matter from particulate to dissolved.

Important in Evolution

It transfers the genes between bacteria, others.

Important model systems in molecular biology.

General Properties of Viruses

Virion

The complete virus particle

It consists of 1 molecule of DNA or RNA enclosed in a coat of protein

It may have additional layers

It cannot be reproduced independent of living cells nor carry out cell division but can exist extracellularly.

Virus Infect cell types

Bacterial Viruses Called Bacteriophages (Phages)

Few archaeal viruses.

Most are eukaryotic viruses.

Plants, animals, protists and fungi.

Classified into families based on genome structure, life cycle, morphology and genetic relatedness.

Virus Structure

Virion size range is ~10–400 nm in diameter and most viruses must be viewed with an electron microscope.

All virions contain a nucleocapsid which is composed of nucleic acid (DNA or RNA) and a protein coat (capsid).

Some viruses consist only of a nucleocapsid, others have additional components.

Capsids

Large macromolecular structures, which serve as protein coat of virus.

Protect viral genetic material and aids in its transfer between host cells.

Made of protein subunits called protomers.

Capsids are helical, icosahedral, or complex.

Helical Capsids

Shaped like hollow tubes with protein walls.

Protomers self assemble.

The size of the capsid is a function of nucleic acid.

Icosahedral Capsids

An icosahedron is a regular polyhedron with 20 equilateral faces and 12 vertices

Capsomers are ring or knob-shaped units made of 5 or 6 protomerspentamers (pentons) – 5 subunit capsomers

hexamers (hexons) - 6 subunits capsomers.

Capsids of Complex Symmetry

Some viruses do not fit into the category of having helical or icosahedral capsids.

poxviruses – largest animal virus.

large bacteriophages – binal symmetry.

the head resembles icosahedral, tail is helical.

Viral Enzymes and Envelopes

Many viruses are bound by an outer, flexible, membranous layer called the envelope

Animal virus envelopes (lipids and carbohydrates) usually arise from host cell plasma or nuclear membranes.

Viral Envelope Proteins

Envelope proteins, which are viral encoded, may project from the envelope surface as spikes or peplomers.

They are involved in viral attachment to host cell

e.g., hemagglutinin of influenza virus

It is used for the identification of a virus

It may have enzymatic or other activity e.g., neuraminidase of influenza virus

It may play a role in nucleic acid replication

Virion Enzymes

It was first erroneously thought that all virions lacked enzymes

Now it is accepted that a variety of virions have enzymes

Some are associated with the envelope or capsid but most are within the capsid.

Viral Genome

Diverse nature of genomes

A virus may have single or double-stranded DNA or RNA

The length of the nucleic acid also varies from virus to virus

Genomes can be segmented or circular.

Viral Multiplication

There are many six steps:

Virus attachment to host cell

Virus entry

Virus uncoating of genome

Virus synthesis

Virus assembly

Virus release

Virus Attachment to the Host Cell

Specific receptor attachment

Receptor determines host preference

It may be specific tissue (tropism)

It may be more than one host

It may be more than one receptor

It may be in lipid rafts providing entry of the virus

Virus Entry and Uncoating

Entire genome or nucleocapsid

Varies between naked or enveloped virus

Three methods used

The fusion of the viral envelope with the host membrane; nucleocapsid Enters

Endocytosis in vesicle; endosome aids in viral uncoating

Injection of nucleic acid

Synthesis Stage

Genome dictates the events.

ds DNA typical flow.

RNA viruses.

Viruses must carry in or synthesize the proteins necessary to complete synthesis.

Stages may occur, e.g., early and late.

Assembly

Late proteins are important in assembly.

Assembly is complicated but varies.

- bacteriophages stages
- some are assembled in the nucleus
- some are assembled in the cytoplasm
- maybe seen as paracrystalline structures in the cell

Virion Release

Nonenveloped viruses lyse the host cell

viral proteins may attack peptidoglycan or membrane

Enveloped viruses use budding

viral proteins are placed into the host membrane

nucleocapsid may bind to viral proteins

envelope derived from the host cell membrane but maybe also from the Golgi, ER, or others

a virus may use host actin tails to propel through the host membrane.

Types of Viral Infection

There are three types of infection:

Infections in *Bacteria* and *Archaea*

Infections in eukaryotic cells

Viruses and cancer

Archaeal and Bacterial Infection

Virulent phage – one reproductive choice

multiplies immediately upon entry

lyses bacterial host cell

Temperate phages have two reproductive options

Either reproduce lytically as virulent phages do

Or remain within the host cell without destroying it

many temperate phages integrate their genome into the host genome (becoming a 'prophage' in a 'lysogenic bacterium') in a relationship called lysogeny.

Lysogenic Conversion

Temperate phage changes phenotype of its host bacteria become immune to superinfection, phage may express pathogenic toxin or enzyme

Two advantages to Lysogeny for Virus

A phage remains viable but may not replicate

the multiplicity of infection ensures the survival of host cell

Under appropriate conditions, infected bacteria will lyse and release phage particles occurs when conditions in the cell cause the prophage to initiate synthesis of new phage particles, a process called induction.

Archaeal Infection

Maybe lytic or temperate

Most discovered so far are temperate by unknown mechanisms

Infection of Eukaryotic Cells

Cytocidal infection results in cell death through lysis.

Persistent infections may last years.

Cytopathic effects (CPEs).

degenerative changes.

Abnormalities.

Transformation to a malignant cell.

Viruses and Cancer

Tumor

growth or lump of tissue;

benign tumors remain in place

Neoplasia

abnormal new cell growth and reproduction due to loss of regulation

Anaplasia

reversion to a more primitive or less differentiated state

Metastasis

the spread of cancerous cells throughout the body

Carcinogenesis

There are two methods:

Complex and multistep process

Often involves oncogenes or cancer-causing genes may come from the virus OR may be transformed host proto-oncogenes (involved in normal regulation of cell growth/differentiation)

Table List of viruses associated with human cancers

Virus	Cancer	Genome type
Hepatitis B virus	Hepatocellular carcinoma	dsDNA
Epstein-Barr virus (EBV)	Several, including Burkitt's lymphoma and nasopharyngeal carcinoma	dsDNA
Human herpesvirus 8 (HHV8)	Several including Kaposi's sarcoma	dsDNA
Human T-cell lympho-tropic virus (HTLV-1)	T-cell leukemia	ssRNA (retrovirus)
Hepatitis C virus	Liver cancer	Single stranded (ss) RNA
Human papillomaviruses strains 6, 11, 16 and 18	Cervical Cancer	dsDNA

Table 2 List of Viruses.

Name of virus	Genome size	Organ effected
Adeno-infection virus or adeno virus(AAV)	26-46 Kbp	Respiratory tract
Aichivirus A	8.2 kb	Gastrointestinal viral diseases

Australian bat lyssavirus/ Pteropid lysavirus	11822 nt	Fatal encephalitis
Astroviridae	6,700 to 7000 nucleotides	Gastrointestinal viral diseases
BK virus	5 kb	Renal transplanta-tion
Banna virus (BAV)	19,500 bp	Mosquito bite
Barmah Forest virus	11,488 nt	Joint inflammation
Bunyamwera orthobunyavi-rus	Tri segmented single-stranded RNA genome L=6875bp S=961bp M=4458bp	Bite of Aedes and Anopheles
La Crosse Bunyavirus	Tri segmented single-stranded RNA genome negative polar-ity L=6980bp S=984bp M=4527bp	------
Snowshoe hare Bunyaviruses	4527 nt	meningoencephalitis
Macacine alphaherpesvirus 1 or herpes virus B	156,789 bp	Lungs, liver PNS to the spinal cord and then the brain
Chandipura vesiculovirus	11 kb	Encephalitis
Chikungunya virus infection	12,000 nucleo-tides long	Exanthematous virus
Cosavirus A (common stool-associated picorna virus)	7-8 kb	--------
Chicken pox /shingles (Varicel-la-zoster)	124884bp	Exanthematous virus
Coxsackie B virus	7,399 nucleotides excluding the 3′ poly A tail	Gastrointestinal to pericarditis and myocarditis
Coxsackievirus A16 (Entero-virus Genus)	7410 nucleotides in length	Hand foot and mouth disease(hfmd)

Crimean-Congo hemorrhagic fever(CCHF)	19.2 kb	Hemorrhagic viral diseases
Coltivirus (Colorado tick fever)	29,000bp	Erythrocytes
Common cold (Rhinovirus)	7200 and 8500 nucleotide	Respiratory tract
Dengue fever	11,0000 bp	Hemorrhagic viral diseases
Dhori virus/ thogotovirus	10 kb	Fever, encephalitis
Dugbe virus disease	18855	Thrombocytes
Duvenhage lyssavirus (DUVV)		Fatal encephalitis
Eastern equine encephalitis (EEEV)	11 kb	encephalaities
Ebola	18-19 kb	Hemorrhagic viral diseases
Echovirus infection	2,179 bp	Gastrointestinal tract
Epstein-Barr virus (EBV)/ Human herpesvirus 4	170 kb	Mononucleosis
European bat 1 lyssavirus (EBLV-1)	11,964 nt	Fatal encephalitis
Fifth diseases (parvovirus B19)	4-6kbp	Exanthematous virus
Orthohantavirus	80-120 nm RNA of three segments.	Renal or respiratory syndrome
Henipa virus/Hendra virus	18.2 kb	Encephalitis
Hepatitis A	7.5 kb	Hepatic viral diseases
Hepatitis B	3,200 bp	Hepatic viral diseases
Hepatitis C	9,600 nucleotides	Hepatic viral diseases
Hepatitis E	7.2 kb	Hepatic viral diseases
Hepatitis D (delta)virus	1.7kb	Hepatic viral diseases
Hepatitis G virus/ GB virus C/ Pegivirus	9.6kb	Hepatic viral diseases
Astroviruses	6,700 to 7,000 nt	Gastrointestinal viral diseases

Coronavirus (alpha)	30 kb	Respiratory tract
Cytomegalovirus(CMV)	236 kbp	Mononucleosis, pneumonia
Enteroviruses	7.2-8.5 kb	Diarrhea, neurological disorder, HFMD
Human Herpes simplex virus 1 (HSV-1)	152 kb	Skin lesions
Human Herpes simplex virus 2 (HSV-2)	125 to 240 kbp depend on species	Skin lesions
Human Herpes simplex virus 6	162,114 bp	Skin lesions
Human Herpes simplex virus 7	145,000 bp	Skin lesions
Human Herpes simplex virus 8	165 to 170 kb	Skin lymphoma
Human Immunodeficiency Virus (HIV)	9.2 kb unspliced genome	White blood cells
Human papillomavirus 1 (HPV)	8 kb	Skin warts
Human papillomavirus 2 (HPV)	8 kb	Skin warts
Human papillomavirus 16,18 (HPV)	7.9 kb and have four region	Cervical cancer, Genital warts
Human parainfluenza(HPIV 2/GREER)	15,654 nt	Respiratory
Human parainfluenza (HPIV2/V98)	15,654nt	respiratory
Parvovirus B19/Primate erythroparvovirus 1	5,596 nt	Skin lesions
Human orthopneumovirus	15 kb	Respiratory
Human Rhinovirus(HRV)	7124 to 7219	Respiratory
Human SARS coronavirus SARS-CoV-2 (beta coronavirus)	26.4 to 31.7 kilobase pairs	Respiratory
Human spumaretrovirus (SFVcpz)	none	Not available

Human T-cell leukemia-lymphoma virus (HTLV-1)	9 kb	Leukemia
Influenza A (H1N1)	13,588 bases long	Respiratory tract
Influenza B	14,548 nucleotides	Respiratory tract
Influenza C	14 kb	Respiratory tract
Isfahan virus/ vesicular stomatitis virus (VSV)	Not available	Undocumented, encephalitis?
Human polyomavirus 2/ JC polyomavirus	5.130 nt	Encephalitis
Japanese encephalitis	11 kb	Encephalitis
Junin virus (JUNV)	L=7.2 kb long S=3.4 kb	Hemorrhagic fever
Human polyomavirus 3/KI virus(KIPyV)	5040 bp	Encephalitis
Kunjin virus Disease	Not available	Encephalitis
Lagos bat lyssavirus (LBV)	NA	Fatal encephalitis
Lake Victoria /Marburg virus	19111 bp	Hemorrhagic fever
Langat virus (LGTV)	10,835 bp	Encephalitis
Lassa fever / Lassa hemorrhagic	A small segment of 3.4 kb and larger segment of 7 kb	Hemorrhagic viral diseases
Lordsdale virus	7555 nt	Gastroenteritis
Louping IIL virus	Approx. 10,800 bp	Encephalitis
Lymphocytic Choriomeningitis Virus (LCMV)	L=7.2kb S=3.4 kb	Encephalitis
Machupo virus	Large 7.2 kb small 3.3 kb	Encephalitis
Mayaro virus (MAYV)	11,429 nt	Fever, joint pain
Measles	15,894 nucleotides	Exanthematous virus
Mengovirus/ Columbia SK virus	7.8 kb	Encephalitis
Merkel cell polyomavirus (MCV350)	5.4 kb	Merkel cell carcinoma

Mokola lyssavirus/ Mokola virus(MOKV)	11,939 nt	Encephalitis
Molluscum contagiosum virus	190 kbp	Cutaneous viral diseases
Monkeypox virus/Orthopoxvirus (MPXY)	197 kb	Cutaneous viral diseases
Mumps	15,384 bp	Mumps
Murray Valley encephalitis virus (MVEV)	11014 bp ss-RNA	Encephalitis
Nipah virus infection	18.2 kb	Respiratory tract
Norwalk virus/ stomach flu/ winter vomiting disease	7.5 to 7.7 kb	Viral Gastroenteritis
O'nyong 'nyong virus (ONNV)	11,800 nt	Fever, joint pain
Orf virus/ Sore mouth infection	138 kbp	Cutaneous viral diseases
Oropouche orthobunyavirus	S=754 bp M=4385bp L=6846bp	Fever, joint pain
Pichinde virus(PICV)	Large 7.2 kb small 3.4 kb	Hemorrhagic fever
Poliovirus	7,000 nucleotides	Neurologic viral diseases
Punta Toro Virus/ bunyaviridae/phelbovirus	S=1.7kb, M= 3.2 kb, L= 6.4kb	Hemorrhagic fever
Puumala orthohantavirus(PUUV)	Large 6.8 to 12 kb medium 3.2 to 4.9 small 1 to 3 kb	Hemorrhagic fever and renal syndrome(HFRS)
Rabies virus	12 kb	Neurologic viral diseases
Rift valley fever (RVF)	11.9 kb	Hemorrhagic fever
Rosavirus	9 kb	
Ross river virus/ Epidemic Poyarthritis	11,892 nt, 11,894 nt,11,894 nt	FEVER, joint pain
Rotavirus A	18,550 bp	Gastroenteritis
Rotavirus B	18,550 bp	Gastroenteritis
Rotavirus C	18,550 bp	Gastroenteritis
Rubella (German measles)	15,894 nucleotides	Exanthematous virus
Sagiyama virus	11,698 nt	Fever, joint pain

Klasservirus/Salivirus A	7633 nt	Gastroenteritis
Sicilian phlebovirus	Not available	Hemorrhagic fever
Sapovirus	747 bp	Gastroenteritis
Semliki forest virus	11442 bp	Fever, joint pain
Seoul orthohantavirus	S=1769bp M=3651bp L=6530bp (clone 1)	Hemorrhagic fever
Simian foamy virus/ spuma-virus	13246bp	None
Simian virus 5 (SV5)	Not available	Undocumented
Sindbis virus(SINV)	11703bp	Fever, joint pain
Southampton virus	Not available	Gastroenteritis
Saint Louis Encephalitis (SLE)	10940 bp	Encephalitis
Powassan virus	10839bp	Encephalitis
Torque teno virus/Transfusion transmitted virus	3852bp	None
Toscana phlebovirus	S=1869bp M=4251 L=6404bp	Hemorrhagic fever
Uukuniemi virus/ Tick-Borne Virus	1720bp M=3229 L=6423bp	Hemorrhagic fever
Vaccinia virus(VACV or VV)	194711bp	None
Variola/small pox	185578bp	Exanthematous virus
Venezuelan equine encephalitis or encephalomyelitis (VEE)	11444bp	Fever, joint pain
Western equine encephalitis virus	11.5 kb	Fever, joint pain
WU polyomavirus(WUPyV)	5229 bp	None
West Nile virus (WNV)	11,000 nucleotide	Mosquito bite
Yaba monkey tumor virus(YMTV)	145,000 bp	None
Yellow fever	11 kb	Hemorrhagic viral diseases
Zika virus	11 kb	Brain and immune system

Name of virus	Genome size	Organ effected
Adeno-infection virus or adeno virus(AAV)	26-46 Kbp	Respiratory tract
Aichivirus A	8.2 kb	Gastrointestinal viral diseases
Australian bat lyssavirus/ Pteropid lysavirus	11822 nt	Fatal encephalitis
Astroviridae	6,700 to 7000 nucleotides	Gastrointestinal viral diseases
BK virus	5 kb	Renal transplantation
Banna virus (BAV)	19,500 bp	Mosquito bite
Barmah Forest virus	11,488 nt	Joint inflammation
Bunyamwera orthobunyavirus	Tri segmented single-stranded RNA genome L=6875bp S=961bp M=4458bp	Bite of Aedes and Anopheles
La Crosse Bunyavirus	Tri segmented single-stranded RNA genome negative polarity L=6980bp S=984bp M=4527bp	------
Snowshoe hare Bunyaviruses	4527 nt	meningoencephalitis
Macacine alphaherpesvirus 1 or herpes virus B	156,789 bp	Lungs, liver PNS to the spinal cord and then the brain
Chandipura vesiculovirus	11 kb	Encephalitis
Chikungunya virus infection	12,000 nucleotides long	Exanthematous virus
Cosavirus A (common stool-associated picorna virus)	7-8 kb	--------
Chicken pox /shingles (Varicella-zoster)	124884bp	Exanthematous virus

Coxsackie B virus	**7,399 nucleotides excluding the 3′ poly A tail**	**Gastrointestinal to pericarditis and myocarditis**
Coxsackievirus A16 (Enterovirus Genus)	**7410 nucleotides in length**	**Hand foot and mouth disease(hfmd)**
Crimean-Congo hemorrhagic fever(CCHF)	**19.2 kb**	**Hemorrhagic viral diseases**
Coltivirus (Colorado tick fever)	**29,000bp**	**Erythrocytes**
Common cold (Rhinovirus)	**7200 and 8500 nucleotide**	**Respiratory tract**
Dengue fever	**11,0000 bp**	**Hemorrhagic viral diseases**
Dhori virus/ thogotovirus	**10 kb**	**Fever, encephalitis**
Dugbe virus disease	**18855**	**Thrombocytes**
Duvenhage lyssavirus (DUVV)		**Fatal encephalitis**
Eastern equine encephalitis (EEEV)	**11 kb**	**encephalaities**
Ebola	**18-19 kb**	**Hemorrhagic viral diseases**
Echovirus infection	**2,179 bp**	**Gastrointestinal tract**
Epstein-Barr virus (EBV)/ Human herpesvirus 4	**170 kb**	**Mononucleosis**
European bat 1 lyssavirus (EBLV-1)	**11,964 nt**	**Fatal encephalitis**
Fifth diseases (parvovirus B19)	**4-6kbp**	**Exanthematous virus**
Orthohantavirus	**80-120 nm RNA of three segments.**	**Renal or respiratory syndrome**
Henipa virus/Hendra virus	**18.2 kb**	**Encephalitis**
Hepatitis A	**7.5 kb**	**Hepatic viral diseases**
Hepatitis B	**3,200 bp**	**Hepatic viral diseases**
Hepatitis C	**9,600 nucleotides**	**Hepatic viral diseases**

Hepatitis E	**7.2 kb**	**Hepatic viral diseases**
Hepatitis D (delta)virus	**1.7kb**	**Hepatic viral diseases**
Hepatitis G virus/ GB virus C/ Pegivirus	**9.6kb**	**Hepatic viral diseases**
Astroviruses	**6,700 to 7,000 nt**	**Gastrointestinal viral diseases**
Coronavirus (alpha)	**30 kb**	**Respiratory tract**
Cytomegalovirus(CMV)	**236 kbp**	**Mononucleosis, pneumonia**
Enteroviruses	**7.2-8.5 kb**	**Diarrhea, neurological disorder, HFMD**
Human Herpes simplex virus 1 (HSV-1)	**152 kb**	**Skin lesions**
Human Herpes simplex virus 2 (HSV-2)	**125 to 240 kbp depend on species**	**Skin lesions**
Human Herpes simplex virus 6	**162,114 bp**	**Skin lesions**
Human Herpes simplex virus 7	**145,000 bp**	**Skin lesions**
Human Herpes simplex virus 8	**165 to 170 kb**	**Skin lymphoma**
Human Immunodeficiency Virus (HIV)	**9.2 kb unspliced genome**	**White blood cells**
Human papillomavirus 1 (HPV)	**8 kb**	**Skin warts**
Human papillomavirus 2 (HPV)	**8 kb**	**Skin warts**
Human papillomavirus 16,18 (HPV)	**7.9 kb and have four region**	**Cervical cancer, Genital warts**
Human parainfluenza(HPIV 2/ GREER)	**15,654 nt**	**Respiratory**
Human parainfluenza (HPIV2/ V98)	**15,654nt**	**respiratory**
Parvovirus B19/Primate erythroparvovirus 1	**5,596 nt**	**Skin lesions**

Human orthopneumovirus	**15 kb**	**Respiratory**
Human Rhinovirus(HRV)	**7124 to 7219**	**Respiratory**
Human SARS coronavirus SARS-CoV-2 (beta coronavirus)	**26.4 to 31.7 kilobase pairs**	**Respiratory**
Human spumaretrovirus (SFVcpz)	**none**	**Not available**
Human T-cell leukemia-lymphoma virus (HTLV-1)	**9 kb**	**Leukemia**
Influenza A (H1N1)	**13,588 bases long**	**Respiratory tract**
Influenza B	**14,548 nucleotides**	**Respiratory tract**
Influenza C	**14 kb**	**Respiratory tract**
Isfahan virus/ vesicular stomatitis virus (VSV)	**Not available**	**Undocumented, encephalitis?**
Human polyomavirus 2/ JC polyomavirus	**5.130 nt**	**Encephalitis**
Japanese encephalitis	**11 kb**	**Encephalitis**
Junin virus (JUNV)	**L=7.2 kb long S=3.4 kb**	**Hemorrhagic fever**
Human polyomavirus 3/KI virus(KIPyV)	**5040 bp**	**Encephalitis**
Kunjin virus Disease	**Not available**	**Encephalitis**
Lagos bat lyssavirus (LBV)	**NA**	**Fatal encephalitis**
Lake Victoria /Marburg virus	**19111 bp**	**Hemorrhagic fever**
Langat virus (LGTV)	**10,835 bp**	**Encephalitis**
Lassa fever / Lassa hemorrhagic	**A small segment of 3.4 kb and larger segment of 7 kb**	**Hemorrhagic viral diseases**
Lordsdale virus	**7555 nt**	**Gastroenteritis**
Louping IIL virus	**Approx. 10,800 bp**	**Encephalitis**
Lymphocytic Choriomeningitis Virus (LCMV)	**L=7.2kb S=3.4 kb**	**Encephalitis**
Machupo virus	**Large 7.2 kb small 3.3 kb**	**Encephalitis**
Mayaro virus (MAYV)	**11,429 nt**	**Fever, joint pain**

Measles	**15,894 nucleotides**	**Exanthematous virus**
Mengovirus/ Columbia SK virus	**7.8 kb**	**Encephalitis**
Merkel cell polyomavirus (MCV350)	**5.4 kb**	**Merkel cell carcinoma**
Mokola lyssavirus/ Mokola virus(MOKV)	**11,939 nt**	**Encephalitis**
Molluscum contagiosum virus	**190 kbp**	**Cutaneous viral diseases**
Monkeypox virus/Orthopoxvirus (MPXY)	**197 kb**	**Cutaneous viral diseases**
Mumps	**15,384 bp**	**Mumps**
Murray Valley encephalitis virus (MVEV)	**11014 bp ss-RNA**	**Encephalitis**
Nipah virus infection	**18.2 kb**	**Respiratory tract**
Norwalk virus/ stomach flu/ winter vomiting disease	**7.5 to 7.7 kb**	**Viral Gastroenteritis**
O'nyong 'nyong virus (ONNV)	**11,800 nt**	**Fever, joint pain**
Orf virus/ Sore mouth infection	**138 kbp**	**Cutaneous viral diseases**
Oropouche orthobunyavirus	**S=754 bp M=4385bp L=6846bp**	**Fever, joint pain**
Pichinde virus(PICV)	**Large 7.2 kb small 3.4 kb**	**Hemorrhagic fever**
Poliovirus	**7,000 nucleotides**	**Neurologic viral diseases**
Punta Toro Virus/ bunyaviridae/phelbovirus	**S=1.7kb, M= 3.2 kb, L= 6.4kb**	**Hemorrhagic fever**
Puumala orthohantavirus(PUUV)	**Large 6.8 to 12 kb medium 3.2 to 4.9 small 1 to 3 kb**	**Hemorrhagic fever and renal syndrome(HFRS)**
Rabies virus	**12 kb**	**Neurologic viral diseases**
Rift valley fever (RVF)	**11.9 kb**	**Hemorrhagic fever**
Rosavirus	**9 kb**	

Ross river virus/ Epidemic Poyarthritis	**11,892 nt, 11,894 nt,11,894 nt**	**FEVER, joint pain**
Rotavirus A	**18,550 bp**	**Gastroenteritis**
Rotavirus B	**18,550 bp**	**Gastroenteritis**
Rotavirus C	**18,550 bp**	**Gastroenteritis**
Rubella (German measles)	**15,894 nucleo-tides**	**Exanthematous virus**
Sagiyama virus	**11,698 nt**	**Fever, joint pain**
Klasservirus/Salivirus A	**7633 nt**	**Gastroenteritis**
Sicilian phlebovirus	**Not available**	**Hemorrhagic fever**
Sapovirus	**747 bp**	**Gastroenteritis**
Semliki forest virus	**11442 bp**	**Fever, joint pain**
Seoul orthohantavirus	**S=1769bp M=3651bp L=6530bp (clone 1)**	**Hemorrhagic fever**
Simian foamy virus/ spuma-virus	**13246bp**	**None**
Simian virus 5 (SV5)	**Not available**	**Undocumented**
Sindbis virus(SINV)	**11703bp**	**Fever, joint pain**
Southampton virus	**Not available**	**Gastroenteritis**
Saint Louis Encephalitis (SLE)	**10940 bp**	**Encephalitis**
Powassan virus	**10839bp**	**Encephalitis**
Torque teno virus/Transfusion transmitted virus	**3852bp**	**None**
Toscana phlebovirus	**S=1869bp M=4251 L=6404bp**	**Hemorrhagic fever**
Uukuniemi virus/ Tick-Borne Virus	**1720bp M=3229 L=6423bp**	**Hemorrhagic fever**
Vaccinia virus(VACV or VV)	**194711bp**	**None**
Variola/small pox	**185578bp**	**Exanthematous virus**
Venezuelan equine encephalitis or encephalomyelitis (VEE)	**11444bp**	**Fever, joint pain**
Western equine encephalitis virus	**11.5 kb**	**Fever, joint pain**
WU polyomavirus(WUPyV)	**5229 bp**	**None**

West Nile virus (WNV)	**11,000 nucleotide**	**Mosquito bite**
Yaba monkey tumor virus(YMTV)	**145,000 bp**	**None**
Yellow fever	**11 kb**	**Hemorrhagic viral diseases**
Zika virus	**11 kb**	**Brain and immune system**

Mechanism of viruses as a cause of cancer:

Viral proteins bind to host cell tumor suppressor proteins

Carry oncogene into the cell and insert it into the host genome

Altered cell regulation

Insertion of promoter or enhancer next to a cellular oncogene

Cultivation of Viruses

Requires inoculation of an appropriate living host

Host for Archael and Bacterial Infection

Usually cultivated in broth or agar cultures of suitable, young, actively growing bacteria

Broth cultures lose turbidity as viruses reproduce

Plaques are observed in agar cultures and CMC (Carboxymethyl cellulose).

Host for animal viruses

Tissue (cell) cultures

Cells are infected with a virus (phage)

Viral plaques

A localized area of cellular destruction and lysis that enlarge as the virus replicates

Cytopathic effects (CPEs)

Microscopic or macroscopic

Degenerative changes or abnormalities in host cells and tissues

Embryonated eggs

Quantification of viruses

Direct counting: viral particles count.

Indirect counting: virus observable.

Hemagglutination assay

Plaque assays

Focus forming unit

Immunofluorescence

Confocal microscopy

RT-PCR (Genomic level)

Infectious unit measure (Concentration of virus)

Plaque assays

Dilutions of virus preparation made and plated on the lawn of host cells

Number of plaques counted

results expressed as plaque-forming units (PFU) – plaque-forming units (PFU)

PFU/ml = number of plaques/sample dilution

Measuring biological effect

Infectious dose and lethal dose assays

Determine the smallest amount of virus needed to cause infection (ID) or death (LD) of 50% of exposed host cells or organisms (ID_{50} or LD_{50}).

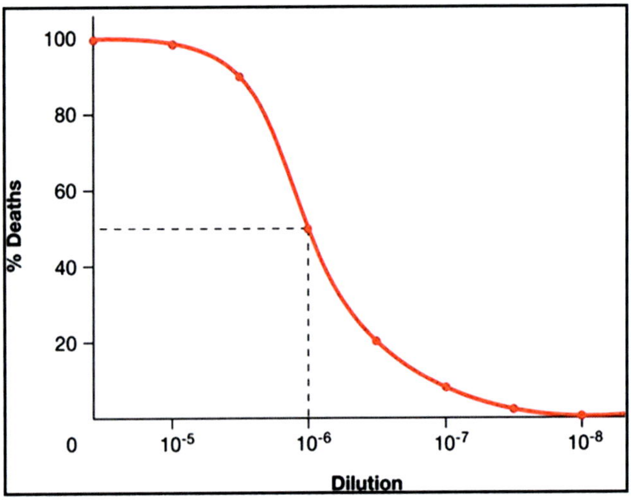

Figure 1.1 Virus titer and percentage of death (Adopted from McGraw Hill companies).

Viroids

Infectious agents composed of closed, circular and ssRNAs

They do not encode gene products and require host cell DNA-dependent RNA polymerase to replicate.

Satellites

Infectious nucleic acids (DNA or RNA).

Satellite viruses encode their own capsid proteins when helped by a helper virus.

Satellite RNAs/DNAs do NOT encode their own capsid proteins.

Encode one or more gene products.

Require a helper virus for replication.

Human hepatitis D virus is satellite.

Prions (Proteinaceous infectious particles)

Cause a variety of degenerative diseases in humans and animals

Scrapie in sheep and bovine spongiform encephalopathy (BSE) or mad cow disease

Creutzfeldt-Jakob disease (CJD) and variant CJD (vCJD) in humans: kuru in humans.

Diseases cause by Prions

- PrP^C (prion protein) is present in "normal" form (an abnormal form of prion protein is PrP^{Sc})
- PrP^{Sc} causes PrP^C protein to change its conformation to the abnormal form
- Newly produced PrP^{Sc} molecules convert more normal molecules to the abnormal form through an unknown mechanism.

Neural Loss

Evidence suggests that PrP^C must be present for neural degeneration to occur

Interaction of PrP^{Sc} with PrP^C may cause PrP^C to crosslink and trigger apoptosis

PrP^C conversion causes neuron loss, PrP^{Sc} is the infectious agent

All prions caused diseases

It has no effective treatment

The result in progressive degeneration of the brain and eventual death

A virus is a particle of DNA or RNA enclosed by structural proteins. Viruses infect cells and proliferate to cause viral infections. Viral skin diseases are classified by clinical features into three types: ① degeneration of epidermal cells and blistering (e.g., in herpes simplex and herpes zoster), ② tumorous changes in epidermal cells (e.g., in verruca vulgaris), and ③ allergic eruptions on the whole body (e.g., in measles and rubella). The first two types are caused by a viral infection in epidermal keratinocytes; the last type is caused by a systemic viral infection (viremia). This chapter introduces various viral skin diseases, including HIV infection.

Researchers can now identify viruses present using metagenomic analyses. This is achieved by comparing the genetic information from next-generation sequencing of clinical samples to the genomes of all known viruses. These include viruses that infect all branches of life, from humans to plants and bacteria. When a sample contains a previously identified virus, its genetic sequences can show upward of 80 percent similarity to viral sequences in public databases such as the National Center for Biotechnology

Information and the European Nucleotide Archive. Such similarities are easily identified computationally.

More challenging are novel viruses whose DNA or RNA genome does not show a significant match to that of any known viruses. In these cases, researchers can translate viral genes into proteins in silico and computationally search for related viral protein sequences.

Due to the redundancy of the genetic code and the need to maintain basic protein structures and active sites, protein sequences evolve at a slower rate than their genes and are therefore recognizable over a longer evolutionary time. With this new ability to rapidly characterize viral genomes, data acquisition is outpacing our understanding of the viruses' role in health and disease. A few years ago, only two polyomaviruses were known to infect humans. Using metagenomics approaches, researchers have identified 13 known human polyomavirus strains, and have linked some of these with diseases ranging from neurological or kidney damage in immunosuppressed transplant and AIDS patients to skin cancers. Most of these polyomaviruses infect a majority of people during childhood and are then silently carried until a weakened immune system unleashes them to wreak havoc. Such occasional pathogenicity is typical of viral families found in humans. For example, some human papillomaviruses are found on the skin of most healthy adults and go unnoticed,[2] while a few specific papillomaviruses can induce cervical or anal cancers (now preventable by early vaccination). Similarly, herpesviruses are nearly universal infections in adults, where they set up lifelong, symptom-free residence in neurons or cells of the immune system. Later in life or following immunosuppression, latent herpesviruses can reactivate and induce diseases ranging from cold sores to meningitis, lymphomas, or Kaposi's sarcoma.

Viral Genome Organization:

A General Overview

All biological organisms have a genome. As we have seen previously, the genome can be either DNA or RNA. This nucleic acid is used to encode functions necessary for it to complete its life cycle and its interaction with its environments. There is great variation in the nature of these genomes. Genome sequencing projects are uncovering many unique features of these that had been previously known.

General features

Currently, over 4000 viruses have been described. These are classified into 71 families. Although viruses are generally the smallest genomes, as a collection of biological genomes they exhibit the greatest variation. The major difference is that some of the genomes are DNA whereas others are RNA. In addition, both DNA and RNA genomes can be either double- or single-stranded (ds or ss). Finally, some ssRNA viruses use the RNA present in the genome to encode genes.

These are called positive-strand ssRNA genomes. Alternatively, negative-strand ssRNA genomes must be copied, and the copy (or negative strand) is used for transcription.

DNA and RNA viral genomes have several distinguishing features. First viral genomes can be monopartite or multipartite. If they are multipartite, they can have several segments. All dsDNA genomes sequenced to date contain only a single nucleic acid molecule. A few of the ssDNA genomes have multiple segments. In contrast, multipartite genomes are much more frequent for RNA viruses. In particular, the negative strand ssRNA viral genomes are generally multipartite.

DNA viruses tend to be larger in size than RNA viruses. It is hypothesized that single- stranded viruses are smaller because that type of molecule is more fragile than the double-stranded molecule. This is generally true for both ssDNA and ssRNA viruses. Some ssDNA genomes can be as small as 1300 nt in length, whereas the minimal genome size of a sequenced ssRNA virus is 2300 nt. Some ssRNA viruses are as large as 31,000 nt, though. The fact that RNA viruses are more susceptible to mutation is thought to have driven these to smaller genome sizes. The largest viruses are the dsDNA viruses, which can be as large as 305,000 nt.

DNA VIRUSES

DNA viruses are typically classified as either 'small' or 'large' genomes (Fig. 1). As the term implies, the difference between these is the size of the genome. Typical of all viruses, these genomes use nearly all of the sequences for genes that encode proteins. The only difference is the actual number of proteins. SV40 (Fig. 2) is a good example of how a small genome can be extensively utilized. Both strands of the 5243 nt dsDNA contain genes. Five of the six genes overlap with another gene. One genomic region is important for the early events of the development of new SV40 virions.

The two early genes are encoded from the negative strand. These genes are actually represented in a single mRNA that is alternatively spliced to encode the small and large T-antigens, proteins critical for launching the replication of the genome. Alternatively, splicing is also key for the late genes that encode the structural components of the virus. In this case, VP2 and VP3 are products of the alternative splicing of the same mRNA. Finally, the VP1 gene overlaps the VP2/VP3 gene region. This feature of maximizing genomic information through gene clustering is a common feature of 'small' genome viruses. Another feature that distinguishes the 'small' and 'large' genomes is the manner in which they are replicated. The 'small' genomes use a host DNA polymerase for replication.

The genome typically encodes proteins that prepare the DNA for replication. In contrast, the 'large' genomes encode a DNA polymerase that is responsible for the genome's replication. As the term implies, 'large' genome DNA viruses can also encode many more proteins than the breadth of genes that can be encoded. The example here is the human adenovirus A. In general, the genome organization is less complex. The plus strand encodes all but five of the genes. Another feature is that few of the genes share sequences with another gene.

RNA VIRUSES

In general, genomes of RNA viruses encode a limited number of proteins. One protein that is very often encoded by these genomes is an RNA-dependent RNA polymerase (RdRp). These polymerases are essential for the replication of both positive and negative strand ssRNAs. as well as dsRNAs. This is true for both monopartite and multipartite RNA viruses that show a range of 1-13 proteins. A major difference between + and – strand ssRNA viruses is that the polymerase is contained within ss (-) virion whereas it is immediately translated from the RNA of the ss(+) RNA.

As with DNA viruses, the genome is an example of maximizing its size. For monopartite ssRNA viruses, the genome encodes a single polyprotein. This protein is then processed into a number of small molecules, each of them critical for the completion of the life cycle of the virus. In multipartite ssRNA genomes, each segment normally contains a single gene. Examples of the two human viruses are depicted.

Retroviruses provide another example of how viruses utilize their minimal sizes. All retroviruses contain *gag*, *pol*, and *env* genes that, respectively, encode structural proteins, the reverse transcriptase, and proteins embedded in the viral coat. These genes generally exist as polyproteins that are processed into several protein products. The reverse transcriptase is the most unique gene for retroviruses. It converts RNA into a DNA copy. This DNA is then used to replicate the genome. In addition, to this basic gene set, retroviruses can encode other genes. The simplest case is the addition of a single gene. Oncogenic retroviruses contain the *gag/pol/env* genes, but also contain a fourth gene called the oncogene. For the Rous Sarcoma Virus, this additional gene is the *src* gene that encodes a tyrosine kinase that modifies protein by adding phosphate groups to tyrosine residues. Many oncogenic retroviruses encode a unique protein that is tied to the cell division process. These genes are similar to other genes found in the host.

The retrovirus version, though, is mutated relative to the host gene. This mutation general leads to uncontrolled cell growth, a phenotypic feature of cancer cells. Human immunodeficiency virus I is the causative agent of AIDS. As with all other retroviruses, it contains the *gag/pol/env* suite of genes. These genes are also expressed as polyproteins that are processed. The additional genes affect other processes including viral infectivity (*vif*), transcription activation (tat), replication (*vp*r, *vpu*, *nef*), and regulation of virion protein expression (*rev*). HIV 1 is an example of a retrovirus that has accumulated multiple genes beyond the basic set. Therefore, it is a good model of retroviral evolution.

Human immunodeficiency virus I (retrovirus)

The HIV I virus is a good example of how a small genome can encode a large amount of genetic information. These eight HIV I genes encode 22 different genes. Three of the genes, gag-pol, gag, and env each encode a polyprotein that is processed into a collection of proteins. It is relatively common for the *gag-po*l, *gag*, and *env* genes of retroviruses to encode multiple proteins. The most important protein for replication of the genome, the reverse transcriptase (RT), is part of the Pol polyprotein. The *gag-po*l polyprotein is cleaved by the protease encoded by the *gag-pol* gene. p2 cleaves the *gag* polyprotein.

One way to make the most from a limited size genome is to use the same sequence for multiple genes. Four of the genes, *ta*t, *rev*, *vpu*, and *nef* each share sequences with the *env* gene. HIV 1 also has a feature in common with other retroviruses. It contains the required gag, pol and genes. It has functions, though, that are defined by a series of other genes. These genes are necessary for viral infectivity (*vif*), transcription activation (*ta*t), replication (*vpr*, *vpu*, *nef*), and regulating virion protein expression (*rev*).

Virus replication

Animals were first used for experimental or diagnostic work, followed by chick embryos and finally cell cultures. Numerous types of animal cell culture have found application in virology. The choice of species, the tissue of origin, and type of culture (primary, cell strain, or cell line) depend on the virus and experimental objectives. Each animal virus can replicate only in a certain range of cells. Among non-susceptible cells, some have a block at an early step. (e.g. they lack suitable receptors or a factor required for expression of viral genes), so that the expression of all viral functions is prevented (resistant cells) Other cells lack a factor required for a later step, so that some, but not all viral activities are expressed (non-permissive cells). In either case, a heterokaryon formed by fusing a susceptible and a non-susceptible cell has the required functions and is usually susceptible.

Role of nucleic acid – transfection

The cells of higher organisms can be infected by naked viral nucleic acid, yielding normal virions. There are several important differences between infections by nucleic acid (transfection) and by virions. **The efficiency of infection with nucleic acid is much lower**, by a factor of 10^{-6} to 10^{-8} in ordinary media, showing the importance of the viral coat in infectivity. The efficiency can be increased by the precipitation of viral DNA onto cells with calcium phosphate, injection or packaging into liposomes.

The host range is much wider with nucleic acids, which can infect resistant cells e.g. chicken cells, although resistant to poliovirus because they lack receptors for the virions, are susceptible to its RNA, but only 1 cycle of viral multiplication takes place because the progeny are again virions and cannot spread to other cells because of the lack of suitable receptors or essential factors.

Infectious nucleic acid can be extracted even from heat- inactivated viruses in which the protein of the capsid has been denatured; the nucleic acid can withstand much higher temperatures than the protein. The ability of nucleic acid infectivity to survive damage to the viral coat must be considered in the preparation of viral vaccines.

With some RNA viruses, a DNA copy of the viral RNA is infectious. e.g. polioviruses, this permits the preparation of viral genomes, such as those of vaccine strains in high quantities by avoiding the high mutation rate in the replication of RNA and its lability.

The infectivity of nucleic acid is unaffected by virus-specific antibodiess, which suggests that this form of a virus could be an effective infectious agent even in the presence of immunity. However, nucleases in the body probably greatly limit its role. Of the animal viruses, papovaviruses, adenoviruses, some herpesviruses, togaviruses and picornaviruses yield infectious nucleic acid. With retroviruses, infectious DNA can be extracted from infected cells or can be made by copying the viral RNA *in vitro*.

Steps in Viral Replication

The following steps take place during viral replication;-

- Adsorption
- Penetration
- Uncoating
- Viral genome replication
- Maturation
- Release

Adsorption

The virus becomes attached to the cells, and at this stage, it can be recovered in the infectious form without cell lysis by procedures that either destroy the receptors or weaken their bonds to the virions. Animal viruses have specialized attachment sites distributed over the surface of the virion e.g. orthomyxoviruses and paramyxoviruses attach through glycoprotein spikes, and adenoviruses attach through the penton fibers. Adsorption occurs to specific cellular receptors. Some receptors are glycoproteins, others are phospholipids or glycolipids. These are usually macromolecules with specific physiological functions, such as complement receptors for EBV. Whether or not receptors for a certain virus are present on a cell depends on the species, the tissue and its physiological state. Cells lacking specific receptors are resistant. Attachment is blocked by antibodies that bind to the viral or cellular sites involved.

Penetration

Penetration rapidly follows adsorption, and the virus can no longer be recovered from the intact cell. The most common mechanism is receptor-mediated endocytosis, the process by which many hormones and toxins enter cells. The virion is endocytosed and contained within a cytoplasmic vacuole.

Uncoating

A key step in uncoating is the acidification of the content of the endosome to a pH of about 5, owing to the activity of a proton pump present in the membrane. The low pH causes rearrangement of coat components, which then exposes normally hidden hydrophobic sites. They bind to the lipid bilayer of the membrane, causing the extrusion of the viral core into the cytosol. For the influenza virus, the acid-sensitive component is the core HA_2 unit of the haemagglutinin, for adenoviruses, it is the penton base.

Viral Nucleic Acid Replication

Virulent viruses, either DNA and RNA, shut off cellular protein synthesis and disaggregate cellular polyribosomes, favoring a shift to viral synthesis. The mechanism of protein synthesis shut-off varies even within the same viral family. Poliovirus, using a viral protease, causes cleavage of a 200 kDa cap-binding protein, which is required for initiation of translation of capped cellular messengers. In contrast to virulent viruses, moderate viruses e.g. polyomaviruses may stimulate the synthesis of host DNA, mRNA, and protein. This phenomenon is of considerable interest in viral carcinogenesis.

DNA virus replication

The mode of replication is semiconservative but the nature of the replicative intermediates depends on the manner of replication. Several methods of replication can be recognized.

A. Adenoviruses - Adenoviruses show asymmetric replication, which initiates at the 3' end of one of the strands using a protein primer. The growing strand displaces the preexisting strand of the same polarity and builds a complete duplex molecule. The displaced strand in turn replicates in a similar manner after generating a panhandle structure by pairing the inverted terminal repetitions.

B. Herpesviruses - Herpesviruses have linear genomes with terminal repeats. On reaching the nucleus, the terminal ends undergo limited exonucleolytic digestion and then pair to form circles. Replication is thought to take place via a rolling circle mechanism, where concatemers are formed. During maturation, unit-length molecules are cut from the concatemers.

C. Papovaviruses - The DNA of papovaviruses is circular and the replication is bidirectional and symmetrical, via cyclic intermediates.

D. Parvoviruses - The replication of single-stranded parvoviruses is initiated when +ve and -ve stranded DNA from different parvovirus particles come together to form a double-stranded DNA molecule from which transcription and replication take place.

E. Poxviruses - The striking feature of poxvirus DNA is that the two complementary strands are joined. The replicative intermediates, present in the cytoplasm, are special concatemers containing pairs of genomes connected either head to head or tail to tail.

F. Hepadnaviruses - Hepatitis B virus employs reverse transcription for replication. The genome consists of a partially double-stranded circular DNA with a complete negative strand and an incomplete positive strand. Upon entering the cell, the positive strand is completed and transcribed. RNA transcripts are in turn reverse-transcribed into DNA by a viral enzyme in several steps, following closely the model of retroviruses, including a jump of the nascent positive strand from one direct repeat (DR) to another.

RNA Viruses

The replication of RNA viral genomes is dictated by the absence of multiple translation units within the same messenger, a characteristic of all animal cell messengers. To overcome this difficulty, 3 main strategies have been developed. The viral mRNA acts directly as the messenger and is translated monocistronically, followed by cleavage to form different proteins. The virion RNA is transcribed to yield various monocistronic mRNAs by initiating transcription at various places. The genome itself is a collection of separate RNA fragments that are transcribed into monocistronic mRNAs. RNA viruses can be placed into 7 classes, according to the nature of the viral RNA and its relation to the messenger.

Class I (e.g. picornaviruses, flaviviruses.) The genome, having +ve polarity, itself acts as the messenger, specifying information for the synthesis of both structural and nonstructural proteins. The same RNA molecule also initiates replication that requires the expression of proteins first. This format allows little control over replication e.g. Poliovirus has no independent mechanism of controlling the numbers of structural proteins made.

Class II (e.g. coronaviruses, togaviruses.) Many +stranded RNA viruses have subgenomic RNA as part of their cycle. This would allow a certain amount of control. The subgenomic mRNA cannot be recognized by the RNA polymerase. It can be used solely for the synthesis of structural proteins etc.

A second way to get around the problem is to make a nested set of RNAs. The nested set of RNA is the most efficient form of control. They can control which part of their genome to express. With togaviruses, the 49S genome RNA is first translated into a polyprotein that is processed into the nonstructural proteins. The subgenomic 26S mRNA, which is transcribed from the full-length -ve RNA, is translated into a smaller polyprotein that is processed into viral structural proteins. With coronaviruses, a nested set of mRNAs is generated in the following manner: the -ve transcript is first generated from the genome, which is then transcribed into monocistronic mRNAs of different sizes. Each begins with an identical short 5' leader sequence that is joined to the transcripts at the start of the various genes and continues to the 3' end of the genome. These mRNAs are not produced by splicing a genomic-size transcript because the virus is able to replicate in enucleated cells.

Class III (e.g. paramyxoviruses, rhabdoviruses.) The genome is of -ve polarity to the messenger. A virion RNA-dependent RNA transcriptase first transcribes the genomes into separate monocistronic messengers initiating at a single promoter. The transcriptase stops and restarts at each juncture between different genes.

Class IV (e.g. orthomyxoviruses, most bunyaviruses.) The -ve genome is in several distinct non-overlapping pieces of ssRNA. The virion transcriptase generates a messenger from each piece. With orthomyxoviruses, most genomic segments contain a single gene but 2 fragments contain 2 overlapping genes: one is expressed by a full-length messenger, the other by a shorter messenger obtained from the former by splicing. The replication of orthomyxoviruses is unusual amongst RNA viruses in that it takes place within the nucleus. The nuclear function it requires is the 5' cap of cellular messengers, which it "pinches" after endonucleolytic cleavage of the host messengers. The 5' cap is then used as primers in the synthesis of viral messengers.

Class V (e.g. arenaviruses, phleboviruses: Arenaviruses have an ambisense genome in that half the genome is of -ve polarity and is transcribed into a messenger by a virion transcriptase, but the other half, which is of +ve polarity is transcribed twice: first a complete transcript of the genome is made, then the mRNA is transcribed from this transcript. This strategy is seen in the S (small) segment of the genome of phleboviruses. Ambisense genomes are unusual for RNA viruses but not for dsDNA viruses.

Class VI (e.g. reoviruses. Reoviruses.) contain distinct non-overlapping segments of dsRNA, each is transcribed into an independent mRNA by the virion transcriptase. Most messengers are monocistronic, but one is bicistronic and expresses a second protein by initiating at an internal AUG in a different reading frame. Each segment of reovirus RNA is replicated independently. A nascent mRNA strand is first generated by the virion transcriptase, which then serves as the template for the replicase to make the negative strand. The two strands remain associated in a dsRNA molecule that ends up in a virion. This replication is asymmetric and conservative because (1) the -ve strand of the virion RNA servers as the initial template and (2) the parental RNA does not end up in the progeny.

Class VII (e.g. retroviruses.) Retroviruses are unique in that their genomes are transcribed into DNA and not RNA. They contain two identical ssRNA of +ve polarity, with a poly-A tail at the 3' end and a cap at the 5' end.

Each is transcribed into DNA by reverse transcriptase that then integrates into the cellular DNA as provirus. Transcription of the provirus by the cellular transcriptase yields the viral molecules that end up in virions. Since RNA viruses of classes III to VII require a virion transcriptase for synthesizing a messenger, their purified viral RNAs are not infectious. Only those of classes I and II are infectious. With RNA viruses, there is no differentiation between early and late messengers.

Replication of Single-Stranded RNA Viruses (Classes I to V)

In all cases, replication consists of building a template strand complementary to the viral strand of the same length, which then servers as the template for progeny viral strands. These steps are carried out by a collection of enzymes of both viral and cellular origin, in association with the nucleocapsids of the infecting virions. In many instances, replication and transcription interfere with each other: with -ve stranded viruses, both template and transcripts are made from viral strands: with +ve viruses, a viral strand can be used as a messenger or replication template. Initially in the infection, there is no interference as the messenger function is needed to provide proteins needed for replication. Later the supply of these proteins regulates the rate of replication. e.g. with poliovirus, replication is initiated when the pVg protein becomes covalently linked to at the 5’ ends of the RNA, apparently initiating the formation of a replication complex. Messenger and progeny often differed structurally. e.g., the messengers of the influenza virus have capped leader sequences derived from cellular messengers.

In addition, they lack 17 to 22 nucleotides at the 3’ end. Moreover, replication requires ongoing protein synthesis to provide the required proteins, whereas transcription does not. In RNA replication, the newly made template strand remains associated with the viral strand on which it is made, forming a double-stranded structure the length of the viral genome, known as the replicative form (RF). Synthesis of new strands occurs by conservative asymmetric synthesis, similar to adenoviruses. An RF with a nascent viral strand is known as RI (replicative intermediate) RF molecules are fairly abundant during replication because, after the completion of a new strand, the replicase appear to remain associated for some time with the template before reinitiating synthesis. RFs accumulate at the end of replication, when no more RIs are formed. With the exception of orthomyxoviruses, the viral RNAs replicate in the cytoplasm. The replicase present in infected cells synthesizes new viral RNA strands of both polarities. Transcription occurs at the same site as replication. It is unclear whether replication and transcription are carried out by different enzymes or by the same enzyme.

Maturation and Release

Maturation proceeds differently for naked, enveloped, and complex viruses.

Naked icosahedral viruses - Preassembled capsomers are joined to form empty capsids (procapsid) which are the precursors of virions. The assembly of capsomers to form the procapsid is often accompanied by extensive reorganization, which is revealed by changes in serological specificity and isoelectric point. eg. picornaviruses and adenoviruses. Naked icosahedral viruses are released from infected cells in different ways. Poliovirus is rapidly released, with death and lysis of infected cells. in contrast, the virions of DNA viruses that tend to mature in the nucleus tend to accumulate within infected cells over a long period and are released when the cell undergoes autolysis, and in some cases, maybe extruded without lysis.

Enveloped Viruses - Viral proteins are first associated with the nucleic acid to form the nucleocapsid, which is then surrounded by an envelope.

In nucleocapsid formation, the proteins are all synthesized on cytoplasmic polysomes and are rapidly assembled into capsid components. In envelope assembly, virus-specified envelope proteins go directly to the appropriate cell membrane (the plasma membrane, the ER, the Golgi apparatus), displacing host proteins. In contrast, carbohydrates and lipids are produced by the host cell. The viral envelope has the lipid constitution of the membrane where its assembly takes place. (eg. the plasma membrane for orthomyxoviruses and paramyxoviruses, the nuclear membrane for herpesviruses) A given virus will differ in its lipids and carbohydrates when grown in different cells, with consequent differences in physical, biological, and antigenic properties.

The envelope glycoproteins are synthesized in the following manner: the polypeptide backbone is first formed on polysomes bound to the ER, which then moves via transport vesicles to the Golgi apparatus where it attains its full glycosylation and fatty acid acylation. The matrix proteins that are present in the viral envelope are usually not glycosylated and stick to the cytoplasmic side of the plasma membrane through hydrophobic domains. Matrix proteins connect the cytoplasmic domains of the envelope glycoproteins with the cell's cytoskeleton, and they gather the viral glycoproteins to form the virions. The selection of viral glycoproteins is efficient but not exclusive. eg. rhabdovirus virions contain 10 to 15% of nonviral glycoproteins. They may also contain glycoproteins specified by another virus infecting the same cell. Envelopes are formed around the nucleocapsids by the budding of cellular membranes.

With orthomyxoviruses and paramyxoviruses, the viral glycoproteins incorporated in the membranes confer on the cell some properties of a giant virion. Thus, cells infected by these viruses may bind RBCs (hemadsoption), and paramyxovirus-infected cells may fuse with uninfected cells to form multinucleated syncytia by the fusion of their membranes. This fusion is equivalent to the fusion of the virion's envelope with the plasma membrane of the host cell at the onset of infection.

Complex Viruses

Maturation of the highly organized poxviruses takes place in cytoplasmic foci called "factories" In contrast to simpler viruses, the poxvirus membrane contains newly synthesized lipids that differ in composition from the cellular lipids. The maturation of poxviruses after the precursors have been enclosed within the primitive membranes suggests that poxviruses may be transitional forms towards a cellular organization.

Defective Interfering Particles

Interference may occur during replication by the generation of defective interfering (DI) particles. They are formed during infection with various kinds of RNA viruses, such as rhabdoviruses, togaviruses, orthomyxoviruses, paramyxoviruses, coronaviruses and some DNA viruses (herpesviruses). With some viruses eg. VSV, the DI particles are smaller than regular particles and can therefore be obtained in pure form. They usually contain the normal virion proteins but have a shorter genome. They are replication-defective and require the helper functions of a normal virus co-infecting the same cell. In early serial passages, DI particles rapidly increase in titer, then the yield of the infectious virus, and finally, the total particle yield is progressively reduced.

The genomes of DI particles are internally deleted but retain both ends, which are essential for the replication of RNA viruses. With DNA viruses, the origin of replication is always conserved and often repeated. Those features show that to cause interference, the DI genomes must replicate. They deprive the regular virus of its replicase by binding to it more effectively. They do not make a replicase of their own because they are always defective in their replicase gene. The formation of DI genomes of RNA viruses is the consequence of the high variability of these genomes. The DI genomes are formed by a copy choice mechanism when the replicase, having replicated part of the template, skips to another part of the same or another template. With VSV and other negatively stranded RNA viruses, 4 types of defective genomes are seen.

- **Deletions** - the polymerase jumps to a site beyond on the same template, skipping a fragment.
- **Snapbacks** - this occurs when the replicase, having transcribed part of the + strand, switches to the just-made - strand as a template. The resultant RNA contains half+ and half- and can produce a hairpin on annealing.
- **Panhandle** - this is formed by a similar mechanism when the polymerase carrying a partially made - strand switches back to transcribing the extreme 5’ of it so that on annealing, the strand forms a panhandle.
- **Compounds** - these genomes are made by a combination of deletions and snapbacks.

- The competition of DI genomes with competent genomes depends not only on the structure of the DI genome but on that of the normal competent genome. Different DI genomes may interfere to very different degrees with the same competent genome and competent genomes may acquire mutations that make them resistant to the existing DI genomes. Subsequently, this is overcome by the new types of DI genomes. During viral multiplication, many types of DI genomes are continuously made and they are very heterogeneous.

Viroids

Viroids are responsible for causing serious diseases in many plants. They consist of naked RNA which does not code for any protein, nor is a protein associated with it. Essentially, each viroid particle is a circular ssRNA molecule containing 250 to 400 nucleotides. They are highly resistant to enzymatic degradation because they have no free ends and because they have a very tight secondary structure All viroid strains have similar characteristics. Their genome can be considered a dsRNA, with many unpaired short "bubbles" regions. There is no AUG initiation codon for protein synthesis, or of their complements (in case the RNA is of negative-stranded type) There is no evidence that the RNA is translated. They are replicated in the nucleus of infected cells by host enzymes through double-stranded intermediates. Replication is blocked by alpha-amantine, which inhibits RNA polymerase II (the RNA polymerase responsible for generating the transcripts for mRNAs).

The base sequences of viroids have repeats, both direct and inverted, which suggest a relatedness to transposing elements. Moreover, they possess a sequence similar to that used by retroviruses. However, viroids are not transcribed into DNA, and no sequences homologous to viroids are found in the DNA of infected cells. cDNA of the viroid is also infectious and can be transcribed into regular infectious viroid particles. A striking feature of viroid RNA is the presence of sequences highly homologous to some of the small nuclear RNAs U_1 and U_3, which are involved in the splicing of introns in animal cells. This suggests that viroids may have originated from introns and their pathogenicity might be due to interference with the normal splicing of introns in cells. Virusoids are satellites of certain plant viruses that are encapsidated with their helper RNAs in the virions. A candidate for a viroid-like agent in humans is the delta agent which is much larger (1678bp) and is surrounded by a coat.

REFERENCES

1. Caspar DLD (1975). Design principles in virus particle construction. In Horsfall FL, Tamm I (eds): Viral and Rickettsial Infections in Man. 4th Ed. JB Lippincott, Philadelphia.
2. Fields BN (ed) (1995). Virology. 3rd Ed. Lippincott-Raven Press.
3. Gajdusek DC (1977). Unconventional viruses and the origin and disappearance of kuru. Science 197:943.
4. Gelderblom HR (1991). Assembly and morphology of HIV: potential effect of structure on viral function. AIDS 5, 617-637.
5. Phil McClean September (2004). Virus Cell Interaction.
6. Mattern CFT (1977). Symmetry in virus architecture. In Nayak DP (ed): Molecular Biology of Animal Viruses. Marcel Dekker, New York.
7. Morse SS (ed) (1994). The Evolutionary Biology of Viruses. Raven Press, New York.
8. Murphy FA, Fauquet CM, Bishop DHL, et al. (eds) (1995). Virus Taxonomy: Sixth Report of the International Committee on Taxonomy of Viruses. Springer-Verlag, New York.
9. Palmer EL, Martin ML (1988) An Atlas of Mammalian Viruses. CRC Press, Boca Raton.
10. Nermut MV, Stevens AC (eds) (1989). Animal Virus Structure. Elsevier, Amsterdam.

CHAPTER 2

VIRUS CELL INTERACTION

INTRODUCTION

A likely unfavorable result of disease in the individual cell is critical to understanding the effect of viral infection complex tissue and brogan and undoubtedly the entire host. As portrayed in the first area, cell tropism of the virus is controlled by the nearness of suitable cell receptors and much of the time, cell-type explicit transmission factor. Virus infections regularly encode qualities of a gene that balance host cell function for their own advantage and obviously the host has expounded natural barriers to limit viral functions. In this way, the viral and cell factors that impacts the result of the disease are regularly in sensitive equalization and effortlessly moved one way or the other.

Virus infection causes a wide variety of potentiality victorious changes in the many different kinds of cells that occur in the animal host. The disruption of cellular function, the induction of cell death or transformation or the activation of an inappropriate immune response are all potentialy manifested as disease by the infected host. Although virus-induced changes at the cellular, sub-cellular and molecular level are most commonly studied in cultured cells, additional insight has been gained through the use of organ culture, transcription of infected cells and tissues back into experimental animals, and the experiment uses of genetically modified laboratory animal in conjunction with molecular clones of individual viruses.

Many virus-cell interactions are transient and asynchronous, involving intermediate steps on the pathway, each featuring dynamic interactions between different viral and cellular components. What could be a better approach to study viral infection than to take a ride with the virus particle on its journey into the cell? To realize this goal, we are developing advanced real-time imaging methods to track the behavior of individual virus particles and their interactions with cellular machinery in live cells. This allows us to follow the fate of individual viruses, to capture fleeting interactions between viruses and cellular components, to dissect these interactions into microscopic steps, and ultimately to determine the molecular mechanisms underlying each step. Our current research is primarily focused on two viruses, influenza and polio, representing enveloped and non-enveloped viruses, respectively.

TYPES OF VIRUS-CELL INTERACTION

A viral infection might be cytocidal or non-cytocidal and beneficial or non-gainful that is, not all disease prompts cell demise or the creation and arrival of new forms. How-ever basic changes can happen in virus-infected cells whether or not the disease is beneficial or not. Probably the most significant of all useless infection cell connections are those related to determined diseases or *persistent infections* or *latent infections*. The term constant contamination just portrays a disease that endures quite a while. The term inert disease portrays contamination that "exists however isn't shown," i.e., a disease wherein irresistible virions isn't framed. In either case, the infection or its genome is kept up uncertainly in the cell, either by the combination of the viral nucleic corrosive into the host cell DNA or via carriage of the viral nucleic corrosive as an episome. In these occasions, the cell endures, to be sure may separate over and over. In certain cases such cells never discharge virions, in others, the disease may get beneficial when instigated by a proper upgrade. Persistent or latent infections may likewise be related to cell change. The different sorts of cooperation that can happen among viruses and cells are summed up.

Cytocidal Changes in Virus-Infected Cells

Cytopathic viruses kill the cells in which they replicate. When a monolayer of cultured cells is inoculated with a cytopathic virus, the first round of infection yields progeny virus that spreads through the medium to infect adjacent as well as distant cells eventually all cells in the culture may become infected. The resulting cell damage is known as a *cytopathic effect* (CPE). The cytopathic effect can usually be observed by low-power light microscopy of unstained cell cultures. The nature of the cytopathic effect is often characteristic of the particular virus involved and is therefore an important preliminary clue in the identification of clinical isolates (see Chapter 12). Fixation and staining of infected cell monolayers may reveal further diagnostic details.

Mechanisms of Cell Damage

So many pathophysiologic changes occur in cells infected with cytopathic viruses that the death of the cell usually cannot be attributed to any particular event; rather, cell death may be the final result of the cumulative action of many insults. Nevertheless, in recent years several specific mechanisms have been discovered, some of which are becoming targets of therapeutic drugs. Particular viruses can cause host cell damage by many different means.

Inhibition of Host Cell Nucleic Acid Synthesis

Inhibition of host cell DNA synthesis is common in viral infections. It is an inevitable consequence of viral inhibition of host cell protein synthesis and its effect on the machinery of DNA replication, but some viruses employ more specific mechanisms. For example, poxviruses produce a DNAse that degrades cellular DNA, and herpes viruses specifically displace the synthesis of host cell DNA with their own synthetic processes.

Inhibition of Host Cell RNA Transcription

Many different classes of viruses, including poxviruses, rhabdoviruses, reoviruses, paramyxoviruses, and picornaviruses, restrain have cell RNA transcription. In certain occurrences, this restraint might be the circuitous result of viral impacts on host cell protein combination, which diminishes the accessibility of transcription factors required for RNA polymerase. In different occurrences, infections encode explicit record factors to manage the statement of their own qualities and, on certain occasions; these elements tweak the outflow of cell qualities too. For instance, herpesviruses encode proteins that quandary straightforwardly to explicit viral DNA sequences, subsequently managing the record of viral genes.

Inhibition of Processing of Host Cell mRNAs

Many viruses, including vesicular stomatitis viruses, influenza viruses, and herpesviruses, interfere with the splicing of cellular primary mRNA transcripts that are needed to form mature mRNAs. In some instances, spliceosomes are formed, but subsequent catalytic steps are inhibited. For example, a protein synthesized in herpesvirus-infected cells suppresses RNA splicing and leads to reduced amounts of cellular mRNAs and the accumulation of primary mRNA transcripts.

Inhibition of Host Cell Protein Synthesis

The shutdown of host cell protein synthesis, while viral protein synthesis continues, is a characteristic of many virus infections. This shutdown is particularly rapid and profound in picornavirus infections, but it is also pronounced in togavirus, influenzavirus, rhabdovirus, poxvirus, and herpesvirus infections. With some other viruses, the shutdown occurs late in the course of infection and is more gradual, whereas, with noncytocidal viruses, such as pestiviruses, arenaviruses, and retroviruses, there is no shutdown and no cell death. The mechanisms underlying the shutdown of host cell protein synthesis are varied: some are as mentioned earlier, whereas others include (1) the production of viral enzymes that degrade cellular mRNAs, (2) the production of factors that bind to ribosomes and inhibit cellular mRNA translation, and (3) the alteration of the intracellular ionic environment favoring the translation of viral mRNAs over cellular mRNAs. Most importantly, some viral mRNAs simply out-compete for cellular mRNAs for cellular translation machinery by mass action; i.e., the large excess of viral mRNA outcompetes cellular mRNA for host ribosomes. Viral proteins may also inhibit the processing and transport of cellular proteins from the endoplasmic reticulum, and this inhibition may lead to their degradation. This effect is seen in lentivirus and adenovirus infections.

Cytopathic Effects of Toxic Viral Proteins

Large amounts of various viral components may accumulate in the cell late in infection. In the past, it was thought that the cytopathic effect was simply a consequence of the intrinsic toxicity of these proteins, but in recent years most cell damage has been recognized as the supervening of viral replication events on cellular events. Hence, the list of “toxic proteins” has been shortened, but some remain. For example, the toxicity of adenovirus penton and fiber proteins seems direct and independent of adenovirus replication.

Cytopathic Changes Involving Cell Membranes

Cellular membranes participate in many phases of viral replication, from viral attachment and entry to the formation of replication complexes, to virion assembly. Viruses may alter plasma membrane permeability, affect ion exchange and membrane potential, induce the synthesis of new intracellular membranes, and induce the rearrangement of previously existing membranes. A generalized increase in membrane permeability, detected by entry into cells of normally excluded macromolecules or escapes of intracellular molecules, occurs early during picornavirus, alphavirus, reovirus, rhabdovirus, and adenovirus infections. Most importantly, enveloped viruses also direct the insertion of their surface glycoproteins, including fusion proteins, into host cell membranes as part of their budding process, often leading to membrane fusion and syncytium formation.

Cell Membrane Fusion and Syncytium Formation

A conspicuous feature of infection of cell monolayers by lentiviruses, paramyxoviruses, morbilliviruses, pneumoviruses, some herpesviruses, and some other viruses is the production of *syncytia* (Figures 5.2C and 5.3C), which result from the fusion of an infected cell with neighboring infected or uninfected cells. Such multinucleate syncytia may also be seen in the tissues of animals infected with these viruses; for example, in horses fatally infected with the Australian equine morbillivirus, a prominent feature of interstitial pneumonia has been alveolar epithelial syncytia (also called multinucleate giant cells). Such syncytia may represent an important mechanism of viral spread in tissues: fusion bridges may allow subviral entities, such as viral nucleocapsids and nucleic acids, to spread while escaping the effects of host defenses. Cell membrane fusion is mediated by viral fusion proteins or fusion domains.

Non-Cytocidal viruses usually don't kill the cell in which they will replicate. On the contrary, they often cause persistent infection, in which infected cells produce and release virions but overall cellular metabolism is little affected. In many instances, the infected cells even continue to grow and divide. This type of virus-cell interaction is found in cells infected with several RNA viruses.

Effect of Non-Cytocidal Viral Infection on Function of Specialized Cell

Some viruses such as pestiviruses, arenaviruses and retroviruses that do not shut down host cell DNA, RNA or protein synthesis or do not kill their host cell produce important pathophysiologic changes in their host by affecting crucial functions that are neither associated with the integrity of the cells or their basic house-keeping functions.

Inclusion Bodies

A characteristic morphological change in cells infected by a certain virus is the formation of Inclusion Bodies, which may be recognized by light microscopy following fixation and staining. Depending on the virus Inclusion Bodies may be intranuclear or intracytoplasmic, single or multiple, large or small, round or irregular shapes and acidophilic or basophilic.

Polarity of Viral Infection

In the course of their replication viruses belonging to several families of enveloped virus insert their glycoprotein into the plasma membrane of their host cell, this insertion is not a random matter. For example, in reservoir host species, rabies virus is shed from the apical surface of salivary gland epithelial cells where it enters the saliva and is transmitted by a bite. However, lentiviruses bud from basolateral plasma membranes and pass directly from cell to cell or become disseminated through tissue spaces and the bloodstream. Wild-type parainfluenza virus 1 (Sendai virus), which causes respiratory disease in several animal species, buds apically into airways and is transmitted by aerosol, whereas a variant that causes generalized disease buds basolaterally.

Ultrastructural Changes in Virus-Infected Cells

As seen at the higher resolution provided by the electron microscope, specific and nonspecific changes in virus-infected cells are varied and may be dramatic. Early changes in cell structure often are dominated by the proliferation of various cell membranes: for example, herpesviruses cause increased synthesis, even reduplication, of nuclear membranes; flaviviruses cause proliferation of the endoplasmic reticulum; picornaviruses and caliciviruses cause a distinctive proliferation of vesicles in the cytoplasm; many retroviruses cause peculiar fusions of cytoplasmic membranes.

Other ultrastructural changes that are prominent in many viral infections include disruption of cytoskeletal elements, mitochondrial damage, and changes in the density of the cytosol. Late in the course of infection, many cytolytic viruses cause nuclear, organelle, and cytoplasmic rarefaction and/or condensation, with terminal loss of host cell membrane integrity. In many instances the inevitability of cell death is obvious, but in others host cell functional loss is subtle and cannot be attributed to particular ultrastructural pathologic changes. In noncytolytic infections, most functional losses cannot be attributed to damage that is morphologically evident. Specific examples reflecting the range of host cell changes occurring in virus-infection.

Molecular Basis of Virus Cell Interaction and Cell Response

The study of molecular aspects of virus-host cell interaction has led to the discovery of vital cellular processes. The notion that viruses have developed tools to effectively recruit the required cellular machinery for their survival and propagation is the basic rationale behind our working hypothesis. We utilize the hepatitis B virus (HBV) as a model because it became evident that HBV is the prototype of a new virus family with a novel life cycle. Additionally, this virus is associated with a wide range of clinical manifestations, including liver cancer (HCC).

Earliest Phases in Infection

HBV does not infect cultured cells therefore our knowledge on the mechanism of the early stages of virus-cell interaction is rather poor. We have developed a protocol to pharmacologically sensitize cells to HBV infection. We also developed a novel tool to measure virus-cell attachment by light microscopy at a single-cell resolution. We identified the QLDPAF sequence within the HBV surface protein (preS1) as the receptor binding viral domain epitope and a second attachment site that recognizes a distinct receptor on the cell membrane. Our results provide evidence for multivalent HBV attachment with synergistic interplay. This cell system was utilized to show that the HBV regulatory protein named pX plays an essential role in infection. Up to this point, the role of pX in the HBV life cycle was not known.

Transcription: Recently, we revealed a novel and much longer HBV transcript (leRNA). leRNA is the first transcript to be expressed and therefore provides the first evidence for HBV enhancer-1 to be activated immediately after transfection. leRNA might potentially encode a novel fused X-Core 40kDa protein, previously detected in HCCs. Unexpectedly, we found that the majority of the leRNA molecules retain in the nucleus, suggesting that HBV programs an RNA nuclear export mechanism yet to be identified.

TOWARDS PRODUCTION OF A SYNTHETIC VIRUS

For gene therapy, tools need to be developed to deliver genetic information to the target cells in safe and efficient ways. Viruses are important but unsafe delivery vectors. This problem can be minimized by the generation of 'synthetic viruses' namely the different virus components are separately synthesized in heterologous systems and then are being reassembled to the desired final product. To this end, we have generated the different constituents of HBV and confirmed their functionality. An interesting outcome was that the recombinant HBV Core particles freely enter the cells (core transduction).

HBV and HCC

Epidemiological and other studies have established a relationship between the HBV carrier-state and the development of HCC. How HBV induces malignancy is an open question. We have evidence for the role of the virus enhancer and the regulatory protein pX in this process. We hypothesized that the HBV enhancer might have an intrinsic oncogenic function and therefore it might be considered as onco-DNA. Interestingly, in HBV associated HCC p53 is mostly wt. This led to the possibility of functional inactivation of p53 by pX. A number of studies reported by us and others support this possibility. Recently, we have isolated a novel gene XAP-8, whose product interacts with pX. XAP-8 is a nuclear protein that plays a role in chromatin remodeling. Interestingly, the XAP-8 gene was amplified in a number of primary tumors, suggesting XAP-8 is a novel oncogene candidate.

P73-cAbl and Novel Apoptotic Pathways

The HBV enhancer interacts with c-Abl. c-Abl, unlike its transforming variants that induce leukemia, is a poor tyrosine-kinase, but it is activated by DNA-damaging agents and γ-irradiation to induce apoptosis. Previously, we have found that p73 involved in this process as well. c-Abl and p73 are physically associated in cellular extracts through the c-Abl- SH3 domain and the p73 PxxP motif. Studies revealed that p73 is a putative c-Abl substrate both *in vivo* and *in vitro*. Interestingly, phosphorylated p73 translocates to the nuclear matrix and even undergoes exonucleosis via the formation of micronuclei. Furthermore, p73 displays unique nuclear mobility and accumulates at the nuclear poles. The significance of this nuclear molecular dynamics that is regulated by protein and possibly DNA redistribution for the HBV life cycle is currently under study.

A Novel Pathway in p53 Degradation

Recently (in collaboration with J. Lotem and L. Sachs) we have found a novel pathway that regulates p53 stability that involves an enzyme NADH quinone oxidoreductase 1 (NQO1). Inhibition of NQO1 by dicoumarol caused degradation of both endogenous and γ-irradiation-induced p53. NQO1 inhibition also induced p53 degradation and blocked p53-mediated apoptosis in γ-irradiated normal thymocytes and in M1 cells. Interestingly, dicoumarol induced degradation of not only wild-type p53 but also of mutant p53, hinting at an Mdm2 independent p53 degradation mechanism. Preliminary results are in support of this possibility. These results raise the possibility of combining dicoumarol with cytotoxic agents in therapy against cancer cells that express high levels of gain of function p53 mutants.

STAGES OF VIRAL INFECTION

Primary infection occurs when the virus enters the body through different portals. The viruses then enter into the bloodstreams and target different organs, the stage is known as viremia. After entry into the predilected site they start their replication and transmitted to different organs and may shed outside through body secretions, the condition is referred to as secondary infection (infection of brain tissue by encephalitis virus and liver by the hepatitis virus).

Primary Infection- Infection of cells in the general location where the virus enters. Common for example with cold viruses and Diarrhea causing viruses. Many viruses including HIV initially infect cells in the infected area.

Viremia- Virus enters the blood system and can be detected in the bloodstream. Not all viruses do this. Some remain only localized.

Secondary Infection- Infection of other organs of cell types by the virus. Many viruses have a high affinity for specific organs due to the presence of receptors or specific cell metabolic functions. Examples are infection of salivary glands by mumps virus, brain tissue by encephalitis virus, and liver tissue by hepatitis virus.

VARICELLA ZOSTER INFECTION

Primary infection: VZV infects via conjunctiva and upper respiratory tract

Days 0 – 3: Replicates in primary lymph nodes

Days 4 – 6: primary viremia,

Days 6 – 13: a replication in the liver, spleen, other organs.

Day 14: infection of skin and appearance of a rash. Infection of sensory ganglia and the establishment of latent infection.

VIRULENCE

The ability of an infectious agent to cause disease. A relative term in the sense that it depends on the particular host that is infected and the state of that host as well as the site of infection.

Rabies and Ebola are highly virulent.

Some viruses infect animals but are virulent only when the host immune response is suppressed. Examples are certain herpes and hepatitis B virus that become virulent when human hosts are treated with immunosuppressant drugs after organ transplant.

CONCLUSION

Resistance to and recovery from viral infections will depend on the interactions that occur between virus and host. The defenses mounted by the host may act directly on the virus or indirectly on virus replication by altering or killing the infected cell. The non-specific host defenses function early in the encounter with a virus to prevent or limit infection while the specific host defenses function after infection in recovery immunity to subsequent challenges. Although the host defense mechanisms involved in a particular viral infection will vary depending on the virus, dose and portal of entry.

REFERENCES

1. Frederick ,A.Murphi, E Pawl J Gibbs, Marrian (2009).Veterinary Virology. 3rd Edition. California, USA.
2. Pallesen G, Hamilton-Dutoit SJ, Rowe M, Young LS: Expression of Epstein-Barr virus latent gene products in tumour cells of Hodgkin's disease. Lancet 337:320, 1999.
3. Weiss LM, Chen YY, Liu XF, Shibata D: Epstein-Barr virus and Hodgkin's disease. A correlative in situ hybridization and polymerase chain reaction study. Am J Pathol 139:1259, 1991.
4. Shibata D, Tokunaga M, Uemura Y, Sato E, Tanaka **S,** Weiss LM: Association of Epstein-Barr virus with undifferentiated gastric carcinomas with intense lymphoid infiltration. Lymphoepitheliomalike carcinoma. Am J Pathol 139:469, 1991.
5. Imai **S,** Koizumi **S,** Sugiura M, Tokunaga M, Uemura Y, Yamamoto N, Tanaka **S,** Sato E, Osato Gastric carcinoma: Monoclonal epithelial malignant cells expressing Epstein-Barr virus latent infection protein. Proc Natl Acad Sci USA 91:913l, 1994.
6. McClain KL, Leach CT, Jenson HB, Joshi VV, Pollock BH, Parmley RT, Dicarlo FJ, Chadwick EG, Murphy SB: Association of Epstein-Barr virus with leiomyosarcomas in young people with AIDS. N Engl J Med 332:12, 1995.
7. Cureton, D.K., Harbison, C.E., Cocucci1, E., Parrish.C.R, and Kirchhausen, T. (2012) Limited receptor crosslinking promotes rapid diffusion of canine parvovirus into clathrin endocytic structures. Journal of Virology 86:5330-5340.
8. Harbison, C.E., Chiorini, J.A., and Parrish, C.R. (2008) The parvovirus capsid odyssey: from the cell surface to the nucleus. Trends in Microbiology 16:208-214.
9. Hafenstein, S., Palermo, L.M., Kostyuchenko, V.A., Xiao, C., Morais, M.C., Nelson, C.D., Bowman, V.D., Battisti, A.J., Chipman, P.R., Parrish, C.R., Rossmann, M.G.(2007). Asymmetric binding of transferrin receptor to parvovirus capsids. Proc Natl Acad Sci U S A. 104: 6585-9.
10. Palermo, L.M., Hafenstein, S.L., Parrish, C.R. (2006). Purified feline and canine transferrin receptors reveal complex interactions with the capsids of canine and feline parvoviruses that correspond to their host ranges. Journal of Virology 80:8482-92.

CHAPTER 3

PESTE DES PETITS RUMINANTS (PPR) VIRUS

INTRODUCTION

Peste des Petits Ruminants (PPR) is an acute highly contagious and fatal disease small ruminants. (Anderson and McKay, 1994; Kulkarni et al., 1996; Raj et al., 2008; Shaila et al., 1989; Shaila et al., 1996). It leads to loss of production particularly in small ruminants and in turn increases poverty in some of the poorest parts of the world. (Anderson and McKay, 1994; Nanda et al., 1996; Shaila et al., 1996). This disease was first reported in West Africa during 1942 (Sen et al., 2010) and now it is wide-spread in tropical and subtropical countries particularly Africa, Middle East and Western, Southern Asia (Abd El-Rahim et al., 2010; Abubakar et al., 2009; Amjad et al., 1996; Bao et al., 2008; Bao et al., 2011b; Cam et al., 2005; Chan, 2010; Couacy-Hymann et al., 2005; Jagtap et al., 2012; Kwiatek et al., 2007; Shaila et al., 1989; Shaila et al., 1996; Taylor et al., 2002). In India, PPR was first recorded in Tamilnadu in 1987 (Shaila et al., 1989) and later found in southern states of Andhra Pradesh, Karnataka, Maharashtra and currently, the disease is endemic in the whole country (Sen et al., 2010). Outbreaks of PPR occur in sheep and goats regularly throughout the year in India, (Balamurugan et al., 2011; Balamurugan et al., 2010a, 2012c; Balamurugan et al., 2010b; Balamurugan et al., 2010c; Couacy-Hymann et al., 2005; Singh et al., 2009; Singh, 2011).

PPR is characterized by pyrexia, ocular and nasal discharge, an erosive lesion on different mucous membranes particularly in the mouth, stomatitis and enteritis leading to foul-smelling diarrhea that ultimately results in death (Gibbs et al., 1979; Lefevre and Diallo, 1990). PPR leads to high morbidity (up to 100%) and high mortality (up to 100%) in sheep and goats resulting in high economic losses. (Abu Elzein and Al-Naeem, 2009; Abu Elzein et al., 1990; Abubakar et al., 2009). Transmission of PPRV mainly occurs during close contact. Inhalation is thought to be an important route of spread. The incubation period can range from 2-10 days; in most cases, signs appear in 2-6 days (Gibbs et al., 1979; Lefevre and Diallo, 1990).

PPR is caused by Peste-des-Petits Ruminant Virus (PPRV) belongs to the genus *Morbillivirus* of the family *Paramyxoviridae*. (Gibbs et al., 1979; Lefevre and Diallo, 1990). PPRV is an enveloped non-segmented negative single-stranded RNA virus (Sen et al., 2010). *Morbillivirus* has six structural proteins: the nucleocapsid protein (NP), the phosphoprotein (P), the large protein (L), the matrix (M), the fusion protein (F), and the hemagglutinin protein (H). PPRV is antigenically closely related to the Rinderpest virus (Gibbs et al., 1979).

For the control of PPR, a live attenuated cell culture vaccine is available (Sen et al., 2010). The heterologous Rinderpest vaccine was used for many years to control PPR but following eradication of Rinderpest this heterologous vaccine was banned because the vaccinated animals could have transmitted the virus to in-contact susceptible animals (Sen et al., 2010). PPRV vaccine is a heat-sensitive, a serious drawback for the efficient use of the live attenuated vaccine in the endemic areas with the hot climate where it is difficult to maintain the cold chain to ensure the potency of the vaccine (Goris et al., 2008; Sen et al., 2010). To minimize the impact of the disease, controlling the spread of the virus is of utmost importance. The vaccine suffers from a set of drawbacks, not the least of which is the time needed to trigger the immune response (immunity-gap). Effective, rapid control tools are, therefore, urgently needed and antiviral compounds could serve this purpose.

There are some antiviral medications approved by the US Food and Drug Administration (FDA) that are used for both prophylactic and therapeutic treatment for viral infection in human beings. However, most of the currently available antiviral agents target viral components repeated use of which leads to the emergence of drug-resistant virus variants due to mutations (Bloom et al., 2010; Fry and Gubareva, 2012; Hamelin et al., 2010; Hayden, 2009;

Hayden, 2006; Hayden and de Jong, 2011; Hayden and Hay, 1992; Ismail et al., 2012; Pawlotsky, 2012; Ujike et al., 2010) which highlights the need of an alternative strategy to design a suitable antiviral compound that does not have a tendency to easily induce drug resistance due to preexisting selection pressure. The cellular factors that are required for virus replication but at the same time are dispensable for host cell metabolism may be a much better target for antiviral interventions as the virus cannot easily replace cellular functions by mutations (Ehrhardt et al., 2010; Eierhoff et al., 2009; Kumar et al., 2011a; Ludwig, 2011; Ludwig et al., 2006; Pleschka et al., 2001).

Living systems respond to the environmental stimuli by encoding and transmitting the received information in form of a chain of events called signal transduction that in turn results in a change in the behavior of the cell. The best-characterized host cell signaling pathway for virus infection are receptor tyrosine kinase (RTK) (Kumar et al., 2011a; Kumar et al., 2011b), Mitogen-activated protein kinase (MAPK) (Droebner et al., 2011; Ludwig et al., 2003; Luig et al., 2010; Planz et al., 2001; Pleschka et al., 2001), phosphatidylinositol 3-kinase (PI3K) (Ehrhardt et al., 2006; Ehrhardt et al., 2007; Hrincius et al., 2011; Marjuki et al., 2006) and Nuclear factor-kappaB (NF-κB) (Flory et al., 2000; Nimmerjahn et al., 2004; Schmolke et al., 2009; Wurzer et al., 2004). Following virus infection, the basic outcome of the activation of such host cell signaling pathways is to produce an antiviral state by secretion of proinflammatory cytokines (Ludwig et al., 2006). However, inhibition of these pathways using small molecule inhibitors or siRNA knockdown of individual host protein (signaling component) have been shown to inhibit virus replication suggesting that the virus misuses the components of signaling pathways to support its own replication. Therefore the virus supportive activity of such signaling pathways might be exploited to develop novel antiviral therapeutics.

Receptor tyrosine kinases are a group of growth factor receptors that, upon ligand binding, undergoes autophosphorylation at Tyr residue. These phosphorylated tyrosines then recruit *Src* homology-2 (SH2) and phosphotyrosine–binding (PTB) domain–containing proteins that activate or link to downstream signaling pathways such as the Ras/ERK/MAPK, PI3K/Akt, JAK/STAT pathways (Band et al., 1997; Biarc et al., 2011; Braunger et al., 1997; Flicker et al., 1997; Grimes and Miettinen, 2003; Hader et al., 2000; Kamakura et al., 1999; Krystal, 2005; Li, 2005; Mayer, 2012; Pauli et al., 2008; Pawson, 1995a, b) (Fig. 1A). There are 58 known RTKs, which fall into 20 subfamilies (Lemmon and Schlessinger, 2010). All

RTKs have a similar molecular architecture, with ligand-binding domains in the extracellular region, a single transmembrane helix, and a cytoplasmic region that contains the protein tyrosine kinase (TK) domain plus additional carboxy (C-) terminal and juxtamembrane regulatory regions (Lemmon and Schlessinger, 2010).

The RTKs have been extensively used as a target for the development of anticancer therapeutics (Hopfner et al., 2006; Ireton and Chen, 2005; Mackiewicz et al., 2011; Nakachi et al., 2010; Zwick et al., 2001). Recent studies (Kumar et al., 2011a; Kumar et al., 2011b) suggests that the receptor tyrosine kinase inhibitor (RTKI) that specifically block nerve growth factor receptor (NGFR) TrkA (NGFR are among one of the 20 members of RTK family having three isoforms, TrkA, TrkB and TrkC) have broad spectrum antiviral properties that inhibit replication of *Influenzaviruses, Rotavirus, Sendaivirus, Mouse hepatitis virus* and *Herpes Simplex Virus.* Further the authors proposed that unlike use of conventional drugs that target the components of pathogen itself (e.g. Amantadine), use of inhibitors that targets some host cell functions (e.g. RTKIs) do not have any tendency to induce drug resistant viruses. We therefore hypothesize whether RTKs can be used as a host target to develop some novel antiviral therapeutic agents against PPR.

Peste des Petits Ruminants (PPR) is an acute highly contagious and fatal disease of small ruminants. (Anderson and McKay, 1994; Kulkarni et al., 1996; Raj et al., 2008; Shaila et al., 1989; Shaila et al., 1996). It leads to loss of production particularly in small ruminants and in turn increases poverty in some of the poorest parts of the world. (Anderson and McKay, 1994; Diallo, 2006; Nanda et al., 1996; Shaila et al., 1996). This disease was first reported in the Ivory cost in West Africa during 1942 (Sen et al., 2010) and now it is widespread in tropical and subtropical countries (Abubakar et al., 2012; Hammouchi et al., 2012; Lefevre and Diallo, 1990) particularly Africa, Middle East and Western, Southern Asia. (Abubakar et al., 2012; Ayari-Fakhfakh et al., 2011; Balamurugan et al., 2012a; Balamurugan et al., 2011; Balamurugan et al., 2012b; Bao et al., 2011a; Bao et al., 2011b; Bodjo et al., 2008; Couacy-Hymann et al., 2007a; Couacy-Hymann et al., 2007b; De Nardi et al., 2012; Delil et al., 2012; Hammouchi et al., 2012; Hoffmann et al., 2012; Khalafalla et al., 2010; Kinne et al., 2010; Kwiatek et al., 2007; Liu et al., 2012; Luka et al., 2012; Munir et al., 2012; Shaila et al., 1989; Shaila et al., 1996; Singh, 2011; Taylor et al., 2002; Wang et al., 2009; Zahur et al., 2011). In India PPR was first recorded in Tamilnadu in 1987 (Shaila et al., 1989) and later found in southern states of Andhra Pradesh, Karnataka, Maharashtra and currently the disease is endemic in the whole country (Balamurugan et al., 2010c; Bhanuprakash et al., 2008; Hegde et al., 2009; Kulkarni et al., 1996; Malik et al., 2011; Raj et al., 2003; Saravanan et al., 2007; Sen et al., 2010; Singh et al., 2009; Singh et al., 2004). Outbreaks of PPR occur in sheep and goats regularly throughout the year, (Balamurugan et al., 2011; Balamurugan et al., 2010a, 2012c; Balamurugan et al., 2010b; Balamurugan et al., 2010c; Raghavendra et al., 2008; Singh et al., 2009; Singh, 2011).

Goats are more susceptible to PPRV infection than sheep (Gibbs et al., 1979). There were descriptions of a PPR outbreak in a zoological collection in the Arabian Gulf and showed clinical infection in *Gazellinae*, *Caprinae* and *Hippotraginae* and subclinical infection in Nilgai (*Tragelophinae*) (Furley et al., 1987). Cattle are susceptible to subclinical infection of PPRV (Anderson and McKay, 1994). PPR is characterized by pyrexia, ocular and nasal discharge, an erosive lesion on different mucous membranes particularly in the mouth, stomatitis and enteritis leading to foul-smelling diarrhea that ultimately results in death (Gibbs et al., 1979; Lefevre and Diallo, 1990). Generally, the morbidity of PPR infection in sheep and goats ranges from 10 to 83% (Anderson and McKay, 1994), with occasional reports of mortality up to 100% (Roeder et al., 1994). The case–fatality rates are higher in goats (71.6%) compared with sheep (55.4%) (Sen et al., 2010). Transmission of PPRV mainly occurs during close contact. Inhalation is thought to be an important route of spread. The incubation period can range from 2-10 days; in most cases, signs appear in 2-6 days (Gibbs et al., 1979; Lefevre and Diallo, 1990).

Molecular epidemiology of PPRV based on the *F* gene sequence from all over the world has defined the existence of four different lineages of virus (I–IV), of which lineage IV is prevalent in the Indian subcontinent and the other lineages are prevalent in African countries (Sen et al., 2010). Analysis of partial sequence data of the *F* gene has pointed out small variations between PPRV strains and has allowed their grouping into four lineages, which better reflects their geographical origins than the variations of genes of the external glycoprotein H (Diallo et al., 2007). Lineage I circulates in West Africa, lineage II in Nigeria and Cameroon, lineage III in East Africa and lineage IV in Asia. However, recently there was an incursion of lineage IV virus into Africa; this was identified following a large epizootic in Morocco (Sen et al., 2010). Based on the partial *F* gene sequencing, there is a solitary report of lineage III PPRV occurrence in Southern India (Shaila et al., 1996). However, since then there has been no further report of this lineage in India. The importance of lineages is that it helps in monitoring of the virus and tracing the source of outbreaks, and is also used for devising control strategies. The prevalence of a particular lineage would help in choosing a candidate vaccine prototype that is homologous to ensure adequate immunization of a susceptible flock. Vaccination with a heterologous lineage would risk the introduction of a lineage hitherto not present. Thus, lineage identification is an important prerequisite of PPR seroepidemiology and control.

PPR virus, along with measles, rinderpest, and canine distemper is included in the *Morbillivirus* genus under the *Paramyxoviridae* family (Gibbs et al., 1979; Sen et al., 2010). Virions of PPRV are pleomorphic, with a lipid envelope enclosing a ribonucleoprotein core that contains the single-stranded negative-sense RNA genome of 15,948 nucleotides (Diallo et al., 2007) and is encapsidated by the nucleocapsid (*N*) protein. The genome codes for six structural (*N*, *P*, *M*, *F*, *H* and *L*) and two nonstructural (*C* and *V*) proteins in the order of 3′-*N*-*P*(*C*/*V*)- *M*-*F*-*H*-*L*-5′ (Bailey et al., 2005; Mahapatra et al., 2006). The viral envelope is derived from the host cell membrane and is associated with three viral proteins: the matrix protein (M), which is located inside the envelope and serves as a link between the nucleocapsid and the two external viral proteins, the fusion protein (F) and hemagglutinin (H).

The virus binds to the host cell receptor through the H protein, followed by fusion of the viral envelope with the cell membrane of the host cell. The F protein mediates virus fusion to introduce the viral nucleocapsids into the cytoplasm where virus multiplication takes place. The progeny virus is released by budding, during which the virus acquires its envelope. Among the viral proteins, F and H are considered to be very important for the induction of protective host immune response against the virus, and most of the neutralizing antibodies are directed against H. Although N is the most abundant and immunogenic viral protein, it does not induce protective immunity against the virus (Diallo et al., 2007). With globalization and intensification of trade, livestock is increasingly exposed to severe animal diseases caused by a range of viruses, an important among which is PPRV. To minimize the impact of the disease, controlling the spread of the virus is of utmost importance. Though the vaccine is commercially available but suffers from a set of drawbacks, not the least of which is the time needed to trigger the immune response (i.e. “immunity-gap”). Effective, rapid control tools are, therefore, urgently needed and antiviral compounds could serve this purpose. Although limited *in vitro* and *in vivo* research has been performed, encouraging results for FMD suggest that livestock could be protected against infection within 24 h following antiviral treatment and up to 12 h post-infection. Such prophylactic/therapeutic antiviral drugs could complement emergency vaccination in a previously disease-free setting or be applied to treat valuable zoological collections and breeding stocks in endemic and previously disease-free regions alike (Charleston et al., 2011; Goris et al., 2008).

There are some antiviral medications approved by the US Food and Drug Administration (FDA) that are used for both prophylactic and therapeutic treatment for viral infection in human beings. However, most of the currently available antiviral agents target viral components repeated use of which leads to emergence of drug resistant virus variants due to mutations (Bloom et al., 2010; Coen and Whitley, 2011; Crumpacker, 1983; Cuevas et al., 2005; Daikoku and Shiraki, 2007; Darby and Larder, 1992; Drew, 1996; Fischer, 1999; Freifeld and Ostrove, 1991; Fry and Gubareva, 2012; Griffiths, 2009; Hamelin et al., 2010; Hayden, 2009; Hayden, 2006; Hayden and de Jong, 2011; Hayden and Hay, 1992; Hirsch and Schooley, 1989; Hovi et al., 2002; Ismail et al., 2012; Kimberlin et al., 1995a; Kimberlin et al., 1995b; Kimberlin et al., 1995c; Kimberlin and Whitley, 1995, 1996; Locarnini and Bowden, 2010; Pawlotsky, 2012; Pillay and Schinazi, 1998; Richman, 2006; Strasfeld and Chou, 2010; Ujike et al., 2010) which highlightens the need of an alternative strategy to design suitable antiviral compound that do not have tendency to easily induce drug resistance due to preexisting selection pressure. The cellular factors that are required for virus replication but at the same time are dispensable for host cell metabolism may be a much better target for antiviral interventions as the virus can not easily replace cellular functions by mutations (Ehrhardt et al., 2010; Eierhoff et al., 2009; Kumar et al., 2011a; Ludwig, 2011; Ludwig et al., 2006; Pleschka et al., 2001; Prober, 2004; Provencher et al., 2004).

As an obligate intracellular parasite virus has to heavily rely on host cell machinery for its effective replication. While the replication cycle and roles of different viral proteins are well characterized, there is a significant gap of knowledge about the host cell factors required by the virus during its replication. Identification of the host cell factors required for viral replication is of importance to understand the mechanism of the viral life cycle and to find new targets for antiviral interventions. Viral pathogens, particularly RNA viruses evolutes frequently due to mutations and become resistant to available antiviral drugs and preexisting immunity. Targeting host cell factor might have an additional advantage in terms of drug resistance because the virus can not easily replace the missing cellular functions by mutations (Ehrhardt et al., 2010; Eierhoff et al., 2009; Kumar et al., 2011a; Ludwig, 2011; Ludwig et al., 2006; Pleschka et al., 2001; Prober, 2004; Provencher et al., 2004).

Living systems respond to the environmental stimuli by encoding and transmitting the received information in form of a chain of events called signal transduction that in turn results in a change in the behavior of the cell. The signals that are activated upon virus infection to cell might be due to (i) interaction of viral surface proteins with cellular receptors (ii) accumulation of viral proteins or RNA inside a cell and (iii) overload of host cell protein-synthesizing machinery due to viral proteins. In addition, many viral proteins not present in infectious virus but produced during replication in host cell might also activate signaling pathways (Mogensen and Jakobsen, 2001). Cellular components that recognize pathogen-associated molecular pattern (PAMPs) are called pattern recognition receptors (PRRs). Three classes of PRRs that sense the viral infection are known- the Toll-like receptors (TLR) 3, 7 and 8 (Jensen and Thomsen, 2012; Schmolke and Garcia-Sastre, 2010), the retinoic acid-inducible gene-I (RIG-1) (Pichlmair et al., 2006) and Nucleotide-binding domain leucine-rich repeat-containing (NLRs) (Allen et al., 2009).

The TLRs are the best characterized PRRs. TLRs can be expressed either on the cell surface (TLR 1, 2, 4, 5, 6,10 and 11) or inside the endosome (TLR 3, 7, 8 and 9) (Aoshi et al., 2011; Koyama et al., 2008; See and Wark, 2008). Surface-expressed TLRs recognize the surface structure of the pathogen while endosomal TLRs recognize pathogen-associated nucleic acid (Guillot et al., 2005) TLR7/8 recognizes ssRNA (Koyama et al., 2008; Munk et al., 2011; Zhang et al., 2010). TLR3 activates the transcription factor AP-1, NF-κB and interferon regulatory factor (IRF) 3 (IRF3) (Kawai et al., 2007). These factors together activate transcription of IFN-β (Kim et al., 1997). NF-κB and AP-1 also lead to the stimulation of proinflammatory cytokines (Ehrhardt et al., 2010). Likewise, TLR7 signaling induces adaptor protein MyD88 leading to the activation of the transcription factor interferon regulatory factor (IRF) 7 (IRF7), AP-1 and NF-κB and (Kawai et al., 2007).

RIG-I is activated by single-strand 5′-triphosphate RNA (Hornung et al., 2006) which is presents only to certain viral RNAs but not on host cell RNA. Upon binding to viral RNA, the conformation of the RIG-I changes which allows it to interact with the mitochondrial antiviral signaling (MAVS) protein (Schmolke and Garcia-Sastre, 2010). MAVS subsequently activates

IRF3 as well as IRF7 and NF-κB. The activated transcription factors initiate the expression of type I IFN and inflammatory cytokines leading to resistance to infection.

The NLRs are recently discovered intracellular PRRs and play an important role in innate immunity in response to the recognition of various "damaged" self (DAMPs or Danger associated molecular pattern released from host cell due to injury caused by either by a pathogen or by other chemical and physical objects) and nonself (PAMP) molecules (Martinon et al., 2009). Inflammasomes are molecular platforms that allow activation of caspase-1 which is an essential regulator of the inflammatory response through its capacity to process and activate proIL-1beta, proIL-18, and proIL-33 (Kuida et al., 1995). NLR protein 3 (NLRP3) also known as NALP3/Cryopyrin/CIAS1/PYPAF1 (Li et al., 2008; McCall et al., 2008) forms inflammasome complex with the adaptor protein ASC (pycard) and caspase-1, which leads to the catalytic cleavage of the immature ('pro') forms of interleukin 1β (IL-1β), IL-18 and IL-33 preventing the tissue damage and necrosis (Martinon et al., 2009). Ichinohe and colleagues recently demonstrated that the influenza virus activates inflammasomes via its intracellular M2 ion channel (Ichinohe et al., 2010) and it plays an important role in the generation of protective adaptive immunity by recruiting influenza virus-specific CD4 and CD8 T cells via ASC, caspase-1 and IL-1R (Ichinohe et al., 2010).

Upon viral infection an array of signal transduction events are initiated by the cells that might be: (i) antiviral (ii) virus supportive (iii) both antiviral and virus supportive and (iv) no role in virus life cycle. The best-characterized host cell signaling pathway for influenza virus infection are Mitogen-activated protein kinase (MAPK) (Droebner et al., 2011; Ludwig et al., 2003; Luig et al., 2010; Planz et al., 2001; Pleschka et al., 2001), Phosphatidylinositol 3-kinase (PI3K) (Ehrhardt et al., 2006; Ehrhardt et al., 2007; Hrincius et al., 2011; Marjuki et al., 2006) Nuclear factor-kappaB (NF-κB) (Flory et al., 2000; Kumar et al., 2008; Nimmerjahn et al., 2004; Pauli et al., 2008; Schmolke et al., 2009; Wurzer et al., 2004). Following their activation, the basic outcome of these pathways is the upregulation of genes responsible for cytokine production (Demirov et al., 2012; Herold et al., 2012; Ludwig et al., 2001b; Ludwig et al., 1999; Mitzner et al., 2009). However, inhibition of these pathways results in inhibition of virus replication suggesting that the virus misuses the component/s of signaling pathways to support its own replication suggesting that these pathways play both antiviral and virus supportive events during influenza virus replication. The virus supportive activity of such signaling pathways is usually exploited to develop novel antiviral therapeutics (Droebner et al., 2011; Ludwig, 2011; Ludwig et al., 2001a; Ludwig et al., 2003; Ludwig et al., 2006; Ludwig et al., 2004; Luig et al., 2010).

Mitogen-activated protein kinase (MAPK) cascades are important signaling pathways that convert extracellular signals into cellular responses. They regulate proliferation, differentiation, cell activation and immune responses. Four different members, organized in separate cascades have been identified so far: (i) ERK (extra-cellular signal-regulated kinase), (ii) JNK (Jun-N-terminal kinase), (iii) p38 and (iv) ERK5. For each MAPK different isoforms are known. All these enzymes are activated by phosphorylation mediated by an upstream MAPK kinase (MAPKK, MEKs or MKKs). It is induced by extra-cellular agents, including pathogens such as RNA viruses and DNA viruses (Pleschka et al., 2008). ERK activation upon influenza virus infection leads to virus-induced cytokine production and airway inflammation (Mizumura et al., 2003), however at the same time appears to support viral replication by facilitating vRNP export (Marjuki et al., 2007; Pleschka et al., 2001).

In the resting stage, NF-κB heterodimer resides in the cytoplasm in a complex with inhibitory kinase proteins (IκB) and can not enter in the nucleus. Various NF-κB inducing signals ultimately leads to activation of IκB kinase β (IKK-β) which in turn leads to phosphorylation of IκB resulting in ubiquitination and proteosome-mediated degradation of IκB. Following degradation of IκB, NF-κB enters into the nucleus and activates transcription of several genes (Ehrhardt et al., 2010). Using two known inhibitors of NF-κB viz: Bay11-7082 which inhibit phosphorylation of IkB and Pyrrolidine dithiocarbamate (PDTC) which inhibits with ubiquitin-proteosome-mediated degradation of IkB (Hayakawa et al., 2003), Kumar et al., identified that the NF-κB signaling differentially regulated influenza virus RNA synthesis and the NF-κB subunit p65 enhances vRNA synthesis but not cRNA synthesis (Kumar et al., 2008). Highly pathogenic avian influenza virus of H5N1 subtypes in humans and birds leads to bleeding and overproduction of cytokines (cytokine storm) preferentially by attacking endothelial cells. The H5N1-induced overproduction of cytokines depends on functional NF-κB signaling whereas low pathogenic strains are much weaker and less NF-κB dependent (Schmolke et al., 2009). Viruses have also evolved strategies to counteract these responses. The NS1 protein of the influenza virus has been shown to suppress IFN-β production (Ludwig et al., 2006). Recent evidence suggests that the influenza A virus also not only suppresses IFN-β induction but also suppresses type I IFN signaling involving NF-κB dependent induction of Suppressor of Cytokine Signaling-3 (SOCS-3) protein expression, which negatively affects JAK/STAT activation ultimately resulting in impaired antiviral response (Pauli et al., 2008).

Phosphatidylinositol-3-kinase (PI3K) has been shown to be activated in response to dsRNA intermediate and mediates activation of the transcription factor interferon regulatory factor-3 (IRF-3) a protein with antiviral functions (Ehrhardt et al., 2006; Ehrhardt et al., 2007; Hrincius et al., 2011; Marjuki et al., 2006). Inhibition of this PI3K pathway using chemical inhibitors results in a decrease in viral titers due to reduced uptake of virus particles (entry) into the host cell (Ehrhardt et al., 2006). Additionally, with mechanism unknown, inhibition of PI3K pathway has also been reported to inhibit viral RNA synthesis, vRNP export and viral protein synthesis (Shin et al., 2007). On the one hand influenza virus NS1 inhibit the dsRNA-responsive transcription factors, on another hand in activates the PI3K /Akt pathway to suppress the onset of premature virus-induced caspase activation and apoptosis (Ehrhardt et al., 2007) by negatively regulating JNK (c-jun N terminal kinase) pathway via ASK1 (apoptosis signal-regulating kinase 1) (Lu et al., 2010).

Receptor tyrosine kinases are a group of growth factor receptors that, upon ligand binding, undergoes autophosphorylation at Tyr residue. These phosphorylated tyrosines then recruit *Src* homology-2 (SH2) and phosphotyrosine–binding (PTB) domain–containing proteins that activate or link to downstream signaling pathways such as the Ras/ERK/MAPK, PI3K/Akt, JAK/STAT pathways (Pauli et al., 2008; Pawson, 1995a). There are 58 known RTKs, which fall into 20 subfamilies (Lemmon and Schlessinger, 2010 Cell). All RTKs have a similar molecular architecture, with ligand-binding domains in the extracellular region, a single transmembrane helix, and a cytoplasmic region that contains the protein tyrosine kinase (TK) domain plus additional carboxy (C-) terminal and juxtamembrane regulatory regions (Fig. 1A). The overall topology of RTKs, their mechanism of activation, and key components of the intracellular signaling pathways that they trigger are highly conserved in evolution from the nematode *Caenorhabditis elegans* to humans, which is consistent with the key regulatory roles that they play (Lemmon and Schlessinger, 2010).

The members of this family of RT cell-surface receptors have emerged as key regulators of critical cellular processes, such as proliferation and differentiation, cell survival and metabolism, cell migration, and cell-cycle control (Blume-Jensen and Hunter, 2001; Ullrich and Schlessinger, 1990).

Furthermore, numerous diseases result from genetic changes or abnormalities that alter the activity, abundance, cellular distribution, or regulation of RTKs. Mutations in RTKs and aberrant activation of their intracellular signaling pathways have been causally linked to cancers, diabetes, inflammation, severe bone disorders, arteriosclerosis and angiogenesis. These connections have driven the development of a new generation of drugs that block or attenuate RTK activity. Basic investigations into the genetics, cellular biology, biochemistry, and structural biology of these receptors have yielded quite a sophisticated view of how this family of proteins functions. These findings have led to the development of numerous important therapeutics and serve as excellent examples of laboratory-driven translational research. Equally important, clinical analyses of RTKs in disease have provided substantial mechanistic insight (i.e., reverse-translational research), which in turn has allowed further refinement of therapeutic strategies.

The molecular mechanisms underlying activation of RTKs share common themes among the families but differ markedly in their details. Only approximately half of the RTK families are well understood, and completing this picture is a major challenge for the future, one that will surely provide many new lessons about RTK activation and signaling. At the level of the receptor molecules themselves, the importance of direct receptor crosstalk or heterodimerization for signaling specificity remains unclear, as does the exact role played by intracellular trafficking. A quantitative understanding of these issues is another key challenge for the future. Finally, as we appreciate the spatial, temporal, and chemical complexity of the signaling networks controlled by RTKs, it becomes increasingly clear that a quantitative biochemical understanding of the behavior of the systems is crucial for predicting qualitative outcomes. Although genetic and biochemical approaches have defined many signaling components in RTK networks, the next major challenge is to put these components into mechanistic and quantitative contexts. This information can then be used to develop new effective treatments for cancers and other diseases driven by activated RTKs.

Besides their extensive role in cancer, recent studies by Kumar et al., (Kumar et al., 2011a; Kumar et al., 2011b) suggests that the receptor tyrosine kinase inhibitor (RTK) (Fig. 1B) that specifically block nerve growth factor receptor (NGF) TrKA (NGFR are among one of the 20 members of RTK family comprise of TrkA, TrkB and TrkC) have broad-spectrum antiviral properties that inhibit replication of *Influenzaviruses, Rotavirus, Sendaivirus, Mouse hepatitis virus* and *Herpes Simplex Virus (HSV-1).*

Further, the authors proposed that unlike the use of conventional drugs that target the components of the pathogen itself (e.g. Amantadine), the use of inhibitors that targets some host cell functions (e.g. RTKIs) do not have any tendency to induce drug-resistant viruses.

Receptor kinase inhibitor tryphostin AG879 (Fig. 1B), MTT [3-(4,5-dimethyl-2-thiazolyl)-2,5-diphenyl-2H-tetrazolium bromide], DMSO and all other chemicals were purchased from Sigma.

PPRV-Nanakpur-2012 was recently isolated from a PPR outbreak that occurred at Nanakpur, Mathura, UP, by Dr. Naveen Kumar at Central Institute for Research on Goats and been described elsewhere (2011, Annual report CIRG). Vero (African green monkey kidney epithelial) cells were procured from National Center for Cell Sciences, Pune. Vero cells were grown in MEM (Sigma) supplemented with 10% fetal bovine serum (Sigma) and antibiotics (Penicillin and streptomycin 1X, Sigma) (Appendix).

TRYPAN BLUE STAINING FOR CELL COUNTING

Cells were counted in hemocytometer by trypan blue staining (a vital stain). After trypsinization cells were diluted 1:4 in trypan blue stain (Sigma) and a drop of it was placed into the hemocytometer fitted with the coverslip. Live cells were counted in all four corners of hematocytometer under an inverted microscope. The average cell count in all four corners was multiplied with the dilution factor to get the number in cells/ml.

Subculture

A confluent monolayer of Vero cells was trypsinize using 0.25% Trypsin-EDTA (Sigma) with a splitting ration of 1:4 and maintained in MEM supplemented with 10% fetal bovine serum and antibiotics.

Plaque Assay for Determination of Viral Titers

The viral titers were determined by plaque assay on Vero cells. Briefly, Vero cells were grown in 6 well tissue culture plates (Corning Inc.,). The confluent monolayer was then infected with 10 fold serial dilution (in 500 μl volume) of the virus (PPRV-Nanakpur-2012) in triplicates for 1h at 37°C after which the infecting media (serum-free MEM supplemented only with antibiotics) was removed and replaced with agar overlay having an equal volume of 2X L-15 medium and 2X agar (Appendix). The plaques were visible by the 3rd-4th day. On day 4th, the aga-overlay was removed with help of a scalpel and the plaques were stained by crystal violet stain (1% crystal violet and 20% methanol, Appendix). The viral titer in pfu/ml (plaque-forming unit/ml) was determined as follows:

Pfu/ml = (Average number of plaques * dilution factor)/ volume of infecting medium For example, if the number of plaques at 10^{-5} dilution were 10, 7 and 8 (Average = 8.33) the titer would be:

$= (8.33*10^5)/0.5$

$= 1.66 * 10^6$ pfu/ml.

Amplification of PPRV

Vero cells were grown in MEM supplemented with 10% fetal bovine serum and antibiotics in 75 cm^2 tissue culture flask (Corning Inc.,). The confluent monolayer was infected with a low multiplicity of infection (moi) (0.001) of PPRV-Nanakpur-2012 in serum-free MEM supplemented only with antibiotics (3 ml volume) for 1 hr at 37°C after which the infecting media (Appendix) was replaced with serum-free MEM supplemented only with antibiotics. Thereafter, the flasks were incubated at 37°C for 3 days after which the supernatant was harvested, aliquoted in 2 ml cryovials and stored in liquid nitrogen. One aliquot was removed from LN2 and viral titer was determined by plaque assay as described above.

Determination of Cytotoxicity (CC_{50}) of AG879

Vero cells in 96-well plates were treated with 2-fold dilutions of AG879 or dimethyl sulfoxide (DMSO) control, in triplicates, in a total of 100 µl growth medium for 48 h. An amount of 20 µl of freshly made 5 mg/ml MTT [3-(4,5-dimethyl-2-thiazolyl)-2,5-diphenyl-2H-tetrazolium bromide] solution (Appendix) was added to each well, and cells were incubated at 37°C for 5 h. After removal of the medium, 200 µl of DMSO was added to each well to dissolve the purple formazan product and the plates were incubated at 37°C for another 5 min to remove any air bubbles. MTT signals were measured photometrically at an absorbance of 550 nm. The 50% cytotoxic concentration (CC_{50}) was calculated as the concentration needed to reduce cellular viability to 50%.

Determination of Effective Concentration (EC_{50}) of AG879

Vero cells in triplicates were infected with PPRV-Nanakpur-2012 at a multiplicity of infection (MOI) of 0.1 for 1 h, washed three times with phosphate-buffered saline (PBS), and replaced with fresh MEM (supplemented with antibiotics only) containing either DMSO or serial 3-fold dilutions of chemical compounds in duplicates. Viral titers in the supernatants at 24 h post-tinfection (hpi) were determined by plaque assay on Vero cells. The 50% effective concentration (EC_{50}) was calculated as the concentration required to reduce viral yield by 50% in the compound-treated cultures when compared to the viral yield in DMSO treated cultures.

Determination of the Virucidal Activity of AG879

Virus suspensions containing approximately 10^6 PFU of PPRV-Nanakpur-2012 were incubated in a serum-free medium containing either DMSO or serial 5-fold dilutions of AG879 for 1.5 h at 37°C. The mixed samples were chilled at 4°C and diluted by 10^{-4}, 10^{-5}-, and 10^{-6} fold with a serum-free medium before being applied onto Vero cells in 6-well plates for plaque assaying. Plaques were scored at 96 hpi after crystal violet staining, and the results were plotted against the concentrations of compounds used. The 50% inactivating concentration (IC_{50}) was calculated as the concentration required to inactivate the cell-free virions by 50%.

EFFECT OF DISRUPTION OF RTK SIGNALING ON PPRV REPLICATION

Effect of Receptor Tyrosine Kinase Inhibitor AG879 on PPRV Replication

Vero cells were infected in triplicate with PPRV-Nanakpur-2012 at MOI # 0.1 for 1 h, washed with PBS three times, and replaced with fresh MEM (supplemented with antibiotics only) with either vehicle control DMSO or 5 μM AG879. Virus production at different time points was quantified by plaque assay.

Effect of shRNA knockdown of (TrkA) on PPRV Replication

Transfection of shRNA Plasmids

Two different short hairpin RNAs (shRNAs) to knock down TrkA expression were used and have been described before (Kumar et al., 2011). For shRNA knockdown, Vero cells were transfected using Escort IV (Sigma) transfection reagent with either the psiRNA-hH1Luc control plasmid, which expresses luciferase-specific shRNA or the combined psiRNA-hH1TrkA1 and -TrkA3 plasmids.

- Grow the cells at ~50% confluency in 6-well tissue culture plates.
- For each well, dilute 2 μg of the plasmid in 100 μl serum and antibiotic-free MEM.
- For each well, dilute 4 μl Escort IV transfection reagent in 100 μl serum and antibiotic-free MEM.
- Mix diluted plasmid and transfection reagent (step ii and iii) (totals volume 200 μl) and incubate it at room temperature for 30 minutes.
- Wash the cells with serum and antibiotic-free MEM.
- Add 0.8 ml serum and antibiotic-free MEM into each well.
- Dropwise add DNA-Lippsome complex solution from step iv onto the cells.
- After 8h add 2 ml MEM (supplemented with 10% serum without antibiotics)
- Next Day change the media (regular growth media with serum and antibiotics).
- At 72 h post-transfection perform the PPRV infection.

Evaluation of the Transfection Efficiency of Escort IV Transfection Reagent

In order to check the transfection efficiency of the Escort IV transfection reagent, a green fluorescent protein (GFP) expressing plasmid pEGFP (available with Dr. Naveen Kumar at CIRG) was used to transfect Vero cells. The cells were observed under a fluorescent microscope for the level of expression of GFP.

Infection with PPRV

At 72 h posttransfection, cells were infected with PPRV-Nanakpur-2012 at an MOI of # 1. Virus titer in the supernatant was determined at 12 hpi by plaque assay.

Evaluation of Knockdown Efficiency of shRNA Plasmids

To evaluate the knockdown efficiency of shRNA-TrkA plasmids, the transfected (and subsequently PPRV infected) cells were scrapped and subjected for total RNA isolation/cDNA preparation for determination of the level of TrkA mRNA (target gene) and β-actin (house-keeping gene) as described below.

Total RNA Isolation

Cells were scraped from tissue culture flask, collected in ice-cold PBS and centrifuged at 1000 rpm for 5 minutes. The cell pellet was then lysed using 1 ml of Tri reagent (Sigma) for 5 minutes. Thereafter, 0.2 ml of chloroform was added, vortexed vigorously for 15 seconds followed by incubation at room temperature for 10 minutes. To separate the organic phase (containing proteins), interphase (containing DNA) and a colorless upper aqueous phase (containing RNA), centrifuge the tubes at 12,000 rpm for 15 minutes. Transfer the upper aqueous phase into a new tube containing 0.5 ml isopropanol, mix gently and incubate at room temperature for 10 minutes. Finally, the RNA pellet was washed twice with 70% ethanol (centrifugation at 12,000 rpm for 5 minutes), dried and dissolved in 30 µl of RNAse/DNAse free water, and store in a deep freezer till use.

Reverse Transcription PCR (RT-PCR)

The knockdown of the TrkA was evaluated by detecting the level of TrkA mRNA present in the cells transfected with TrkA siRNA expressing plasmids (psiRNAhH1-TrkA) using RT-PCR. The primers used for amplification of TrkA were: Forward primer- 5′-CCC CAG CCA TGT ACG TTG CTA TCC -3′ and reverse primer- 5′-GCC TCA GGG CAG CGG AAC CGC TCA -3′. β-actin was used as a housekeeping control gene and was amplified using forward primer 5′-GAGTCTCTCCTGGAAAACTG-3′ and reverse primer 5′-CATTTGTTGAGCACAAGGAGC-3′. The primers were designed using Primer3 software and analyzed by online software available at IDT DNA.

Reverse Transcription (cDNA Preparation)

The cDNA was synthesized using reagents procured from Fermentas in a total reaction volume of 20 µl as follows:

RNA: 5 µl

100 µM Random hexamer (100uM): 1 µl

RNAse/DNAse free water: 6 µl

Mix, incubate at 65°C for 5 minutes, chill on ice, spin briefly and add

5X RT buffer: 4 μl

10 mM dNTPs: 2 μl

40U/μl RNAse inhibitors: 1 μl

200U/μl RevertAid H Minus Reverse Transcriptase: 1μl

Mix and incubate at 25°C for 10 minutes, 42°C for 1 hr and 70°C for 10 minutes. Store at -20°C.

Polymerase Chain Reaction (PCR)

PCR reaction was carried out in 1 total 50 μl volume containing the followings:

Template (cDNA): 5ul

Forward Primer (100 pmol): 1 μl

Forward Primer (100 pmol): 1 μl

10 mM dNTPs: 2μl

DMSO: 3ul

10X PCR buffer: 5 μl

25mM MgCl2 (Fermentas): 5 μl

Taq DNA polymerase (Fermentas): 1 μl

RNAse/DNAse free water: 27 μl

For amplification of TrkA, initial denaturation of 95°C for 5 minutes followed by 40 cycles of 95°C for 30 seconds, 52°C for 30 seconds and 72°C for 1 minute and a final extension step of 72°C for 10 minutes. For amplification of β-actin, similar thermocycler conditions were used except annealing at 54°C for 30 seconds and extension at 72°C for 30 seconds.

Agarose Gel Electrophoresis

To visualize the PCR amplified DNA fragments of TrkA (target gene) and β-actin (housekeeping gene), PCR products were run into 1% agarose gel (Appendix). Agarose gel was prepared by mixing 1 gm of agarose (Sigma) and 10 μl (10 mg/ml stock) of ethidium bromide into 100 ml of 1X TAE buffer (Appendix) and boiling it till completely dissolved. The gel was poured into the tray fitted with a comb when the temperature of the gel reached about 50-60°C. The gel tray was filled with 1X TAE buffer, the samples were prepared by mixing with loading dye (1:6) (Fermentas) and loaded into the wells using 10 μl microtip. Finally, the gel was run at 100V for 1hr. Amplification of the DNA fragments was visualized and recorded using a gel documentation system. The size of the DNA fragments was compared with 100 bp DNA marker (Banglore Genei) loaded side by side with samples in the gel.

Peste des Petits Ruminants (PPR) is an acute highly contagious viral disease of small ruminants that leads to high morbidity and high mortality. To minimize the impact of the disease, controlling the spread of the virus is of utmost importance. The vaccine is commercially available but suffers from a set of drawbacks, not the least of which is the time needed to trigger the immune response (immunity-gap). Effective, rapid control tools are, therefore, urgently needed and antiviral compounds could serve this purpose. In this study, the anti-PPRV activity of a receptor tyrosine kinase inhibitor tyrphostin AG879 was studied in a wide variety of cell culture-based assays. Further studies by shRNA knockdown of RTK (TrkA) and subsequent PPRV infection confirmed the functional significance of RTK signaling in PPRV replication.

PPRV Life cycle is of 6-8 hrs

Since there is no definitive information in the literature about the time duration of the PPRV life cycle, therefore it was first planned to carry out a time-course experiment where Vero cells were infected with high Moi of PPRV-Nanakpur-2012 (moi # 5) and the virus particles in the supernatant at different time post-infection were quantified by plaque assay (Fig. 2). There was no significant rise in the viral titers in supernatant that were collected at 1 hpi, 2 hpi, 3 hpi and 4 hpi (Fig 3). The detectable amount of virus particles (<1 log) at these earlier time points (1 hpi, 2 hpi, 3 hpi and 4 hpi) represents the background level (input) which remained attached with vessel/cells even after extensive washing following infection. However, there was a sharp rise in viral titers in supernatants that were collected following 6 hpi which indicated that the new progeny virus particles start releasing somewhere around 6 hpi (Fig. 3). The maximum virus particles released between 6-8 hpi after which (8 hpi, 10 hpi, 12 hpi and 24 hpi) there was no major increase in viral titers (almost steady-state curve) (Fig. 3). These findings suggested that the PPRV life cycle is 6-8 hrs.

Determination of the Cytotoxicity (CC_{50}) of Receptor Tyrosine Kinase Inhibitor tryphostin AG879

The CC_{50} (i.e., the concentration required to produce cytotoxic effects in 50% of target cells) was determined by using an MTT assay to estimate the viability of Vero cells grown in the presence of increasing concentrations (up to 40 μM) of AG879. As shown in Fig. 4, no cytotoxicity was observed even after 72 h of incubation of Vero cells with AG879 at 5 μM but significant cell death was observed at >5 μM (CC_{50} ~ 15 μM). Therefore it was decided to use 5 μM of AG879 in subsequent experiments.

Determination of the Virucidal Activity (IC_{50}) of Receptor Tyrosine Kinase Inhibitor AG879

The inactivation concentration 50 (IC_{50}) or virucidal activity (Concentration of AG879 required to inactivate 50% of the extracellular virus) of AG879 on the extracellular virus was determined by incubating the PPRV-Nanakpur-2012 with indicated concentrations of AG879 for 1.5 h and then titrating it on Vero cells. We did not observe any significant inactivation of PPRV-Nanakpur-2012 even at 81 μM of AG879 (Fig. 5) which suggested that the AG879 has no effect on the extracellular virus.

Receptor Tyrosine Kinase Inhibitor Blocks PPRV Replication

In order to determine the anti-PPRV activity of receptor tyrosine kinase inhibitor AG879, Vero cells were either infected with low moi (0.1) or high moi (2) of PPRV-Nanakpur-2012 and the progeny virus released in supernatant was quantitated at 24 hpi and 10 hpi respectively by plaque assay in Vero cell. As compared to vehicle (DMSO) control, the AG879 inhibited the PPRV replication ~ 200 times (>2 log) either it was infected with low moi (Fig. 6A) or high moi (Fig. 6B), though the overall titers in supernatant were more when the cells were infected with low moi (~log 7.5 in low moi and ~log 5 in high moi) (Fig. 6). These findings suggested that the receptor tyrosine kinase signaling is a prerequisite for PPRV replication and may serve as a target for the development of antiviral drugs.

In order to determine the effective inhibitory concentration (EC_{50}) of AG879 on PPRV replication, a dose-response experiment was performed. As depicted in Fig. 7, as compared to vehicle DMSO control, the receptor tyrosine kinase inhibitor, AG879 significantly inhibited the PPRV replication in the μM range (at 1 μM but not at 0.33 μM) (Fig. 7).

shRNA Knockdown of Receptor Tyrosine Kinase (TrkA) leads to Impaired PPRV Replication

In order to confirm whether the inhibitory effect of AG879 on PPRV replication is really due to its action on its target molecule (RTK) and not due to off-target effects, the Vero cells were either transfected with TrkA siRNA expressing plasmids (psiRNAhH1-TrkA) or with negative control siRNA expressing plasmid (psiRNAhH1-Luc). At 3 days post-transfection, when the knockdown of TrkA is believed to occurred (Fig. 8B), the cells were infected with moi # 1 of PPRV-Nanakpur-2012 and the supernatant was collected at 10 hpi to quantitate the progeny virus released in supernatant. The cells that were transfected with TrkA siRNA expressing plasmids produced significantly less virus as compared to negative control siRNA expressing cells (Fig. 8A). These findings with shRNA knockdown of TrkA confirm the functional role of RTK signaling in PPRV replication.

To observe the knockdown efficiency of shRNAs, the level of the expression of TrkA and a housekeeping gene β-actin was evaluated in infected cell lysate by determining the level of respective mRNA by reverse-transcription PCR (RT-PCR). As depicted in Fig 8B, the TrkA expressing shRNA specifically and significantly knockdown the level of TrkA which was not observed with negative control shRNA expressing plasmids. Further, the TrkA shRNA expressing plasmid also did not affect the expression level of the off-targets as the level (mRNA) of housekeeping gene β-actin was similar in both TrkA and negative control shRNA expressing cells (Fig. 8A).

In order to determine the transfection efficiency of Escort IV transfection reagent in Vero cells, a GFP expressing plasmid, pEGFP was used to transfect side-by-side with shRNA expressing plasmids. On fluorescent microscopic examination of pEGFP transfected cells at 3 days post-transfection, about 60% of cells were found to express the GFP (Fig. 8C).

Taken together, the data suggested that the receptor tyrosine kinase signaling is required for PPRV replication and may serve as a target for the development of antiviral drugs.

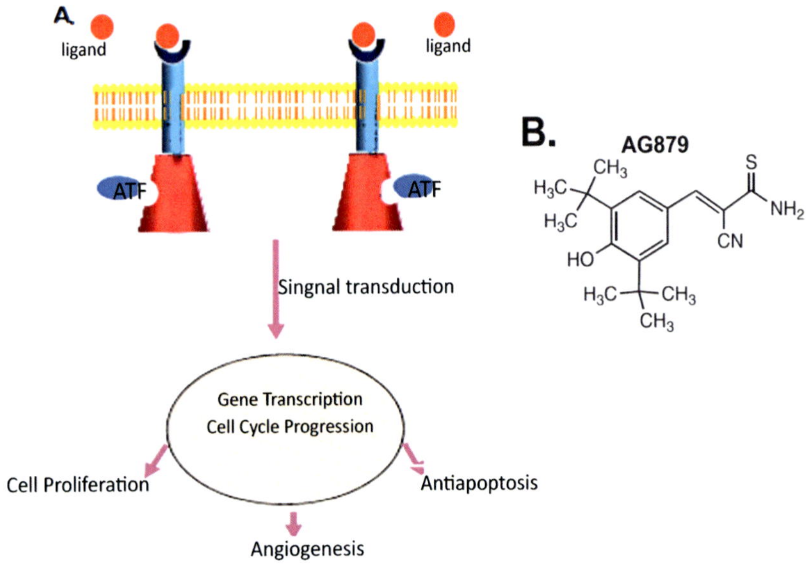

Figure 1. Receptor tyrosine Iciness signaling: (A). Receptor tyratine kinases (RTICs) cue located on plasma membrane On binding of a suitable ligand to edracellular domain, the cytoplasmic domain of the RTK undergoes ataRtskos-siyabgo Lat tyrosine residue) and dtmafzabcn within turn leads to activation of several downstream kinases in a chain of ments called signal transduction. Signal transchwbort result gene transcription and cell cycle regulabon required for cell proliferation, cell survival. angiogenesis and apoptsis. (Adopted from Kutner et al, unpublished) (B). Chemical structure of a receptor tyamine kinase inhibitor torbosito AG879.

Receptor Tyrosine Kinase Inhibitor AG879 Appears to Inhibit Multiple Post-Entry step(s) of PPRV Replication

In order to specifically determine which step(s) of PPRV replication are affected by AG879, a time-course experiment was performed. Vero cells were infected with high moi (moi # 2) and the inhibitor or vehicle control DMSO was applied at different times post-infection. The virus particles in the supernatant were quantified at 10 hpi. As indicated in Fig. 9, as compared to vehicle control DMSO, the AG879 inhibited the PPRV replication almost at a similar level whether it was applied at 1 hpi or 2 hpi or 3 hpi which suggested that very early step(s) such as attachment and entry which are believed to occur within 1-2 hpi, are unlikely to be affected. However, as compared to earlier time points (1-3 hpi), there was a sharp rise in viral titers when the inhibitor was applied at 4 hpi, a time point where RNA synthesis is believed to occur (Fig. 9). As compared to DMSO vehicle control, a marginal though the statistically significant difference in the viral titers was also observed even when the inhibitor was applied at 4 hpi, 5 hpi or 6 hpi which suggests that at a certain extent, the inhibitor also appears to affect the late stage(s) of PPRV replication (Fig 9). Taken together it appears from these primary studies that the receptor tyrosine kinase inhibitor AG879 affects multiple post-entry steps of PPRV replication.

Peste des Petits Ruminants Virus (PPRV) is an important viral pathogen across the developing world that places a huge disease burden on the animal industry particularly in small ruminants and leads to loss of production and in turn, increases poverty in some of the poorest part of the world. PPR leads to high morbidity (10-83%) and high mortality (10-100%) in sheep and goat (Sen et al., 2010) resulting in high economic losses (Khan et al., 2008; Lefevre and Diallo, 1990; Sen et al., 2010).

Due to close homology of PPRV to rinderpest virus, live attenuated tissue culture rinderpest vaccine (TCRV) has been successfully used in past to control PPR but since the eradication of rinderpest, TCRV vaccine has been banned and homologous PPR vaccine has been developed worldwide (Sen et al., 2010). The homologous PPR vaccine is believed to provide protection for at least 3-4 years. However, despite vaccination, the disease is emerging and reemerging probably due to insufficient vaccine coverage and lack of diagnostic tools to facilitate the diagnosis during epidemics. To minimize the impact of the disease, controlling the spread of the virus is of utmost importance. Though vaccines suffer from a set of drawbacks, not the least of which is the time needed to trigger the immune response (i.e. "immunity-gap"). Effective, rapid control tools are, therefore, urgently needed and antiviral compounds could serve this purpose. Although limited *in vitro* and *in vivo* research has been performed, encouraging results for FMD suggest that livestock could be protected against infection within 24 hpi following antiviral treatment and up to 12 hpi. Such prophylactic/therapeutic antiviral drugs could complement emergency vaccination in a previously disease-free setting or be applied to treat valuable zoological collections and breeding stocks in endemic and previously disease-free regions alike (Charleston et al., 2011; Goris et al., 2008).

In this study, using small molecule cell signaling inhibitor tryphostin AG879 and shRNA knockdown we found that a host cell's receptor tyrosine kinase i.e. Nerve Growth Factor Receptor (NGF or TrkA) is required for efficient PPRV replication and hence may be targeted to develop future antiviral agents against PPRV.

Since there is no definitive information available on the PPRV replication cycle which is essentially required to study the effect of drugs at various stage(s) of the viral life cycle, therefore, we first determined the duration of the PPRV life cycle by time-course experiment in Vero cell culture model (Fig. 3). As compared to 1 hpi, 2 hpi and 4 hpi, a sharp increase in the viral titers in the supernatant at 6 hpi indicated the release of new progeny virus particles which means that one cycle of the virus has occurred at this point. There was a sharp increase in the virus titers till 8 hpi after which there was no major increase which suggested that one full cycle has completed in 6-8 hpi. We also did not observe any major increase in the viral titers even at 12 hpi and at 24 hpi, the point at which the second/third/fourth cycle should have occurred to cause a major shift in viral titers. This might be because of the high moi (moi =5) of virus infection used for this experiment (supposed to infect and lyse all possible cells in a single replication cycle). Like influenza and some other viruses, PPRV also causes an acute infection and should have a short replication cycle. We got almost a similar kinetics (6-8 hpi) of PPRV replication like that of influenza virus (7-8 hpi) (Kumar et al., 2011a).

An important step in drug screening in the cell culture model is to first test the cytotoxicity of the individual drug. We did not observe any cytotoxicity at ≤5μM of AG879 (Fig. 4) which suggested that 5μM of AG879 is safe to use for the cell culture (Vero) model. Therefore 5μM of AG879 was used for all subsequent experiments. Likewise, for testing of antiviral compounds in a cell culture model, it is an essential step to see the virucidal effect of the drug on the extracellular virus. We did not observe any virucidal activity of AG879 even at 81μM (Fig. 5) which suggested that the anti-PPR activity of AG879 is via inhibition of its target (RTK) and not simply due to inactivation of the extracellular virus.

Upon viral infection an array of signal transduction events are initiated by the cells that might be: (i) antiviral (ii) virus supportive (iii) both antiviral and virus supportive and (iv) no role in virus life cycle. The best-characterized host cell signaling pathway for influenza virus infection are Mitogen-activated protein kinase (MAPK) (Droebner et al., 2011; Ludwig et al., 2003; Luig et al., 2010; Planz et al., 2001; Pleschka et al., 2001), Phosphatidylinositol 3-kinase (PI3K) (Ehrhardt et al., 2006; Ehrhardt et al., 2007; Hrincius et al., 2011; Marjuki et al., 2006) Nuclear factor-kappaB

(NF-κB) (Flory et al., 2000; Kumar et al., 2008; Nimmerjahn et al., 2004; Pauli et al., 2008; Schmolke et al., 2009; Wurzer et al., 2004). Following their activation, the basic outcome of these pathways is the upregulation of genes responsible for cytokine production (Demirov et al., 2012; Herold et al., 2012; Ludwig et al., 2001b; Ludwig et al., 1999; Mitzner et al., 2009).

However, inhibition of these pathways results in inhibition of virus replication suggesting that the virus misuses the component/s of signaling pathways to support its own replication suggesting that these pathways play both antiviral and virus supportive events during influenza virus replication. The virus supportive activity of such signaling pathways is usually exploited to develop novel antiviral therapeutics (Droebner et al., 2011; Ludwig, 2011; Ludwig et al., 2001a; Ludwig et al., 2003; Ludwig et al., 2006; Ludwig et al., 2004; Luig et al., 2010).

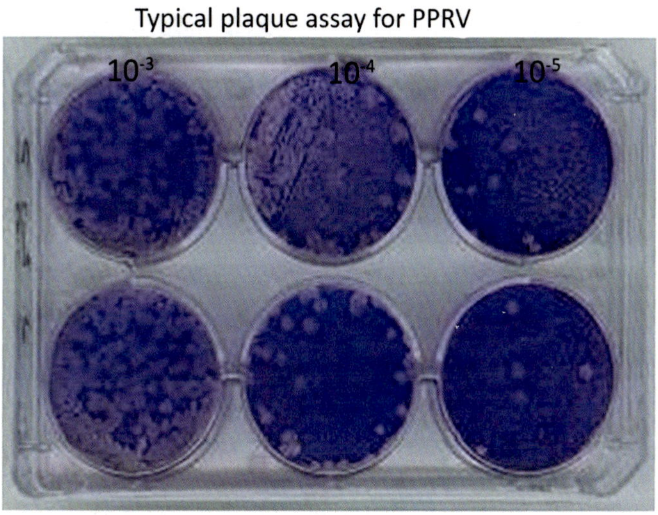

Figure 2. Quantitation of PPRV by plaque assay: Vero cells were grown in 6-well tissue culture plats. Confluent monolayer was then used to infected with serial dilution of PPRV for lh after which the infecting media was replaced with agar-overlay. Hates were further incubated at 37°C for 4 days. The plaques were then stained with crystal violet

Receptor tyrosine kinases are a group of growth factor receptors that, upon ligand binding, undergoes autophosphorylation at Tyr residue. These phosphorylated tyrosines then recruit *Src* homology-2 (SH2) and phosphotyrosine–binding (PTB) domain–containing proteins that activate or link to downstream signaling pathways such as the Ras/ERK/MAPK, PI3K/

Akt, JAK/STAT pathways (Pauli et al., 2008; Pawson, 1995a). There are 58 known RTKs, which fall into 20 subfamilies (Lemmon and Schlessinger, 2010 Cell). All RTKs have a similar molecular architecture, with ligand-binding domains in the extracellular region, a single transmembrane helix, and a cytoplasmic region that contains the protein tyrosine kinase (TK) domain plus additional carboxy (C-) terminal and juxtamembrane regulatory regions. The overall topology of RTKs, their mechanism of activation, and key components of the intracellular signaling pathways that they trigger are highly conserved in evolution from the nematode *Caenorhabditis elegans* to humans, which is consistent with the key regulatory roles that they play (Lemmon and Schlessinger, 2010).

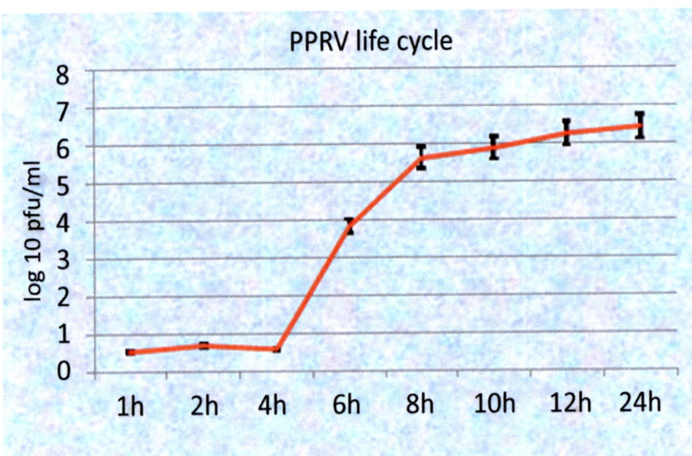

Figure 3. PPRV life cycle is of 6-8 ivy Confluent monolayer of Vero cells were infected with MOI # 5 of PPRV-Nanakpur-2012 in triplicates for 1h at 37°C after which the cells were washed 6 times with increasing volume of PBS and fresh media (MEM supplemented with penicillin and streptomycin) was added. The plate's were then incubated at 37°C. At indicated time points, supernatant was collected and quantified by plaque assay in Vero cells.

The members of this family of RT cell-surface receptors have emerged as key regulators of critical cellular processes, such as proliferation and differentiation, cell survival and metabolism, cell migration, and cell-cycle control (Blume-Jensen and Hunter, 2001; Ullrich and Schlessinger, 1990). Furthermore, numerous diseases result from genetic changes or abnormalities that alter the activity, abundance, cellular distribution, or regulation of RTKs. Mutations in RTKs and aberrant activation of their intracellular signaling pathways have been causally linked to cancers, diabetes, inflammation, severe bone disorders, arteriosclerosis and angiogenesis. These connections have driven the development of a new generation of drugs that block or attenuate RTK activity. Basic investigations into the genetics, cellular biology, biochemistry, and structural biology of these receptors have yielded quite a sophisticated view of how this family of proteins functions. These findings have led to the development of numerous important therapeutics and serve as excellent examples of laboratory-driven translational research. Equally important, clinical analyses of RTKs in disease have provided substantial mechanistic insight (i.e., reverse-translational research), which in turn has allowed further refinement of therapeutic strategies.

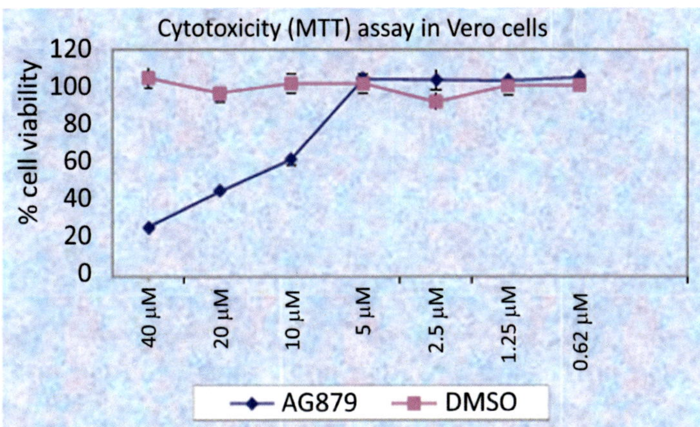

Figure 4. Cytotoxicity (MTT) assay of receptor tyrosine kinase inhibitor tryphostin AG879: Vero cells Were treated with 10-fold dilutions of AG879 or dinethyl *-411code (DM9)) control, in triplicates for 48 It An amount of 20 pl, of freshly made 5 mg/nil MTT solution was added to each well, and cells were incubated at 37°C for 5 h. After removal of the medium, 200 ul of DM90 was added to each well to dissolve the pimple formazan product and the plates were incubated at 37°C for another 5 min to remove any air bubbles. MTT signals were measured photometrically at an absorbance of 550 run The 50% cytotoxic concentration (CC5o) was calculated as the concentration needed to reduce cellular viability to 50 %.

The molecular mechanisms underlying activation of RTKs share common themes among the families but differ markedly in their details. Only approximately half of the RTK families are well understood, and completing this picture is a major challenge for the future, one that will surely provide many new lessons about RTK activation and signaling. At the level of the receptor molecules themselves, the importance of direct receptor crosstalk or heterodimerization for signaling specificity remains unclear, as does the exact role played by intracellular trafficking. A quantitative understanding of these issues is another key challenge for the future. This information can then be used to develop new effective treatments for cancers and other diseases driven by activated RTKs.

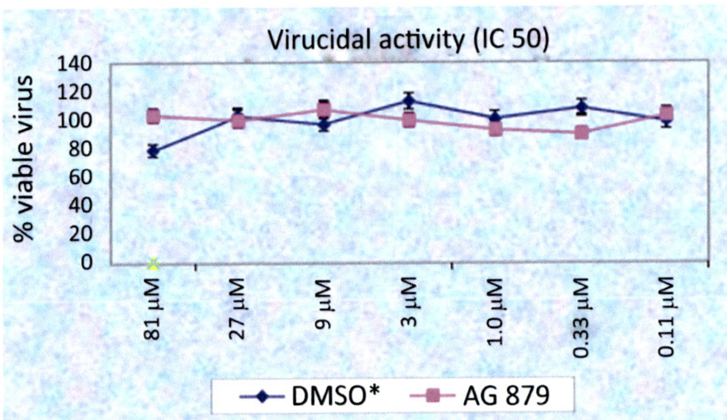

Figure 5. Virucidal activity of receptor tyrosine kinase inhibitor trvphostin AG879: Virus suspensions containing approximately 105 PFU of PPRV-Nanal-cpur-2012 were incubated in serum-free medium containing either DM9) (vehicle control) or serial 3-fold dilutions of AG879 for 1.5 h at 37°C. The inbred samples were chilled at 4°C and diluted by 104, 105, and 106 fold with serum-free medium before being applied onto vero cells in 6-well plates for plaque as_i ja Plaques were scored at 96 hpi after crystal violet staining, and the results plotted against the concentrations of compounds used The 50% inactivating concentration (IC5o) was calculated as the concetration required to inactivate the cell-free virions by 50%.

Besides their extensive role in cancer, recent studies by Kumar et al., (Kumar et al., 2011a; Kumar et al., 2011b) suggests that the receptor tyrosine kinase inhibitor (RTK) that specifically block nerve growth factor receptor (NGF) TrKA (NGFR are among one of the 20 members of RTK family comprise of TrkA, TrkB and TrkC) have broad-spectrum antiviral properties that inhibit replication of *Influenzaviruses, Rotavirus, Sendai virus, Mouse hepatitis virus* and *Herpes Simplex Virus (HSV-1)*. Further, the authors proposed that unlike the use of conventional drugs that target the components of the pathogen itself (e.g. Amantadine), the use of inhibitors that targets some host cell functions (e.g. RTKIs) does not have any tendency to induce drug-resistant viruses.

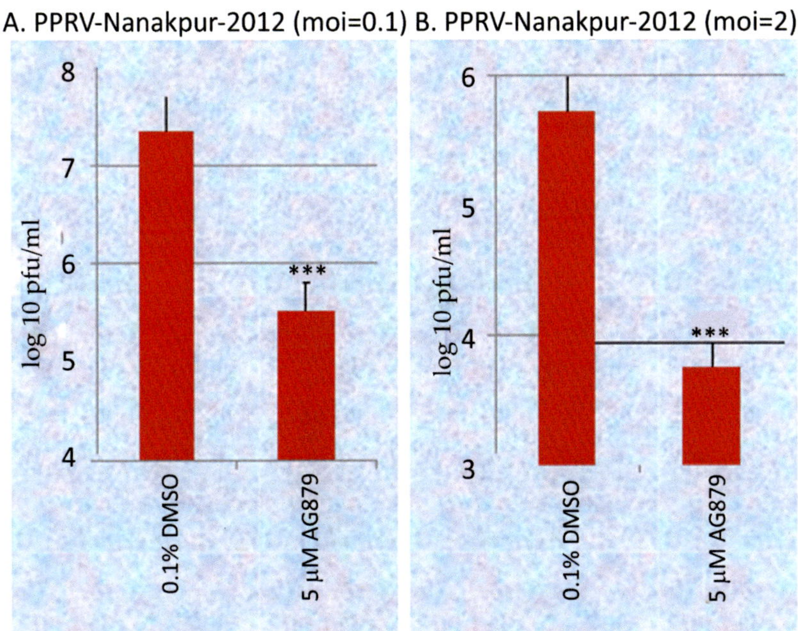

Figure 6. Effect of receptor tyrosine kinase inhibitor AGS79 on PPRV replication: (A). Confluent monolayer of Vero cells were infected in triplicates with PPRV-Nanakpur-2012 at moi of 0.1, washed 3 times with PBS, added with fresh media (MEM supplemented with antibiotics only) having 5 of AG879 or 01 % DYED (vehicle control) and incubated for 48 his . The supernatant was quantified by plaque assay in Vero cells. (B). Confluent monolayer of Vero cells were infected in triplicates with PPRV-Nanakpur-2012 at moi of 2, washed 6 times with PBS, added with fresh media (MEM supplemented with antibiotics only) haying 5 pM of AGS79 or 01% DM50 (vehicle control) and incubated for 12 his . The supernatant was quantified by plaque assay in Vero cells. m= P<0.001.

In this study, we have shown the small molecule cell signaling inhibitor tryphostin AG879 that is known to block RTK signaling can potently block PPRV replication in µM range underlying varying dose of virus infection (high and low moi) (Fig. 6 and 7). AG879 is a tyrphostin-class compound that selectively antagonize the TrkA/HER2 pathway. By testing specific shRNA knockdown, we were able to verify the importance of RTK (TrkA) in PPRV replication. These findings serve to extend prior studies that have also suggested important roles for RTK signaling in the influenza virus life cycle using genome-wide screens conducted to search for host factors involved in virus replication (Brass et al., 2009; Cohen-Daniel et al., 2009; Hao et al., 2008; Karlas et al., 2010; Konig et al., 2010; Watanabe et al., 2010). siRNA knockdown of RTKs in these studies resulted in reduced virus replication, supporting the functional role of RTK signaling in virus replication. Thus, RTK signaling may be required for the PPRV life cycle and so may present novel targets for antiviral drug development. Though beyond the scope of this study but further studies are required to demonstrate the kinetics of TrkA phosphorylation (and its inhibition via AG879) following PPRV replication.

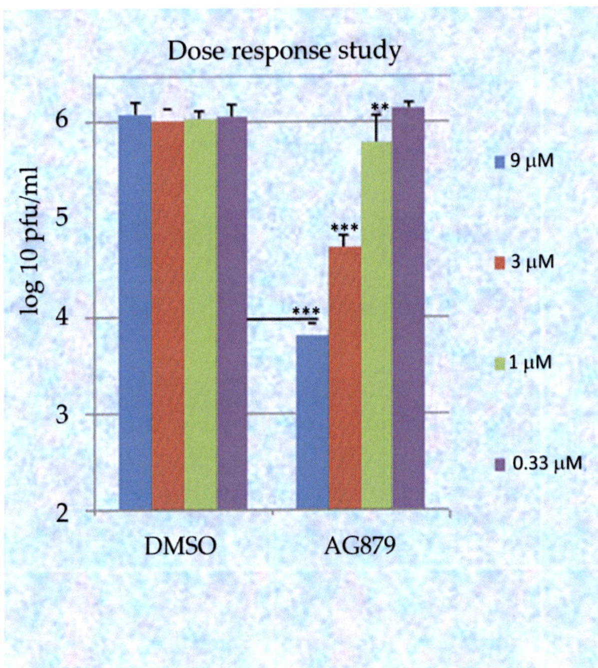

Figure 7: Dose response study (EC50) of receptor tyrosine kinase inhibitor AG879 on PPRV replication: (A). Confluent monolayer of Vero cells were infected in triplicates with PPRII-Nanakpur-2012 at moi of 0.1., washed 3 times with PBS, added with fresh media (MEM supplemented with antibiotics only) hating indicated concentration of AG879 or equal volume of DMS0 (halide control) and incubated for 48 his. The supernatant was quantified by plaque assay in Vern cells. m= P<0.001, *"=P<0.01.

Further, we examined the stage (s) of the PPRV life cycle affected by RTKI, AG879. It appears that the RTK inhibitor AG879 block multiple post-entry steps of the PPRV life cycle (Fig. 9). It is unlikely that this RTKI would inhibit early step(s) of PPRV replication as the magnitude of the inhibition of the viral titers were almost similar (as compared to DMSO vehicle control) when the inhibitor was applied either at 1 hpi or at 2hpi or at 3 hpi, the expected time points of early steps of the viral life cycle (attachment, entry, uncoating). We did not observe any major inhibition in viral titers when the inhibitor was applied at 4 hpi (Fig. 9) which suggested that the post-entry step(s) of the PPRV life cycle (e.g. RNA and protein synthesis) are most likely to be affected. However a statistically significant, though the lower level of inhibition of viral titers was also observed when the inhibitor was applied at later time points (4 hpi, 5 hpi and 6 hpi) which suggested that this AG879 also block late stages of PPRV replication. It's beyond the scope of this study to examine the effect of RTKIs on individual steps of PPRV replication to more precisely make a statement which step(s) of the viral life cycle is actually affected. Likewise, the molecular mechanism of interaction between RTK and its downstream signaling components to the virus counterpart is also unexplored and needs to be investigated in the future. Nevertheless, similar kinetics of the inhibition of viral titers has also been reported recently in influenza virus replication, though being an RNA virus it replicates in the nucleus (Kumar et al., 2011a; Kumar et al., 2008).

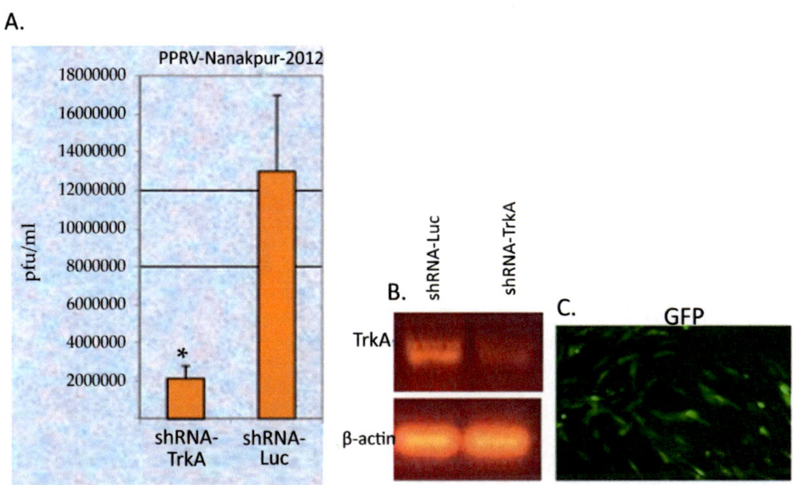

Figure 8. shRNA knockdown of RTK (TrkA) leads to impaired PPRV replication. Vero cells were transfected with either TrkA siRNA expressing plasmids (psiRNAhH1-TrkA1/3) or negative control siRNA expressing plastid (psiRNAhl-n-Luc). Three days following transfection, cells were infected with PPRV- Nanakpur-2012 atmoi, 1. Supernatant was collected a t 10 l for quantitation of the virus (A) Cells were scrapped and subjected for total RNA prepara-

tion/cENA synthesis b determine the level of vaRNA (TrkA and β-actin) by PCR (B). Transfection efficiency of Escort JV transfection reagent as determined by transfection of a GFP expressing plasmid, pEGFP at3 days-post infection under florescent microscopy (C). *= P<0.05.

The RTKIs inhibitors tryphostin AG879 and tryphostin A9 have been shown to block multiple post-entry steps (RNA synthesis, vRNP export from nucleus to cytoplasm and budding) of the influenza virus replication (Kumar et al., 2011a). To the extent that any given drug might target multiple host components and multiple steps of virus replication, drug-resistant viral variants are less likely to occur (Kumar et al., 2011b)(Kumar et al., 2011).

As indicated above the AG879 appears to affect multiple post-entry steps of PPRV replication (Fig. 9) so there are fewer possibilities of emergence of drug-resistant variants of PPRV under the selective pressure of AG879. Further, as compared viral components as a target, drugs that target host components (e.g. blocking RTK signaling using AG879 here) is less likely to produce drugresistant virus variants because the virus can not replace missing cellular functions by mutations. However, it is worth noting that, given their low therapeutic indexes *in vitro*, AG879 is unlikely to be useful as antiviral agents themselves, but nonetheless they point the way to developing or discovering better antiviral drugs.

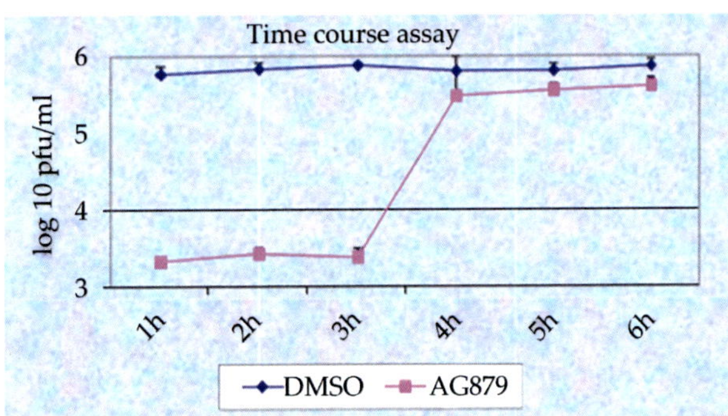

Figure 9. Receptor tyrosine kinase inhibitor AG879 appears to affect multiple post-entry steps of PPRV replication: (A). Confluent monolayer of Vero cells were infected in triplicates with PPRV-Nanakpur-2012 at moi of 2, wasted 6 times with PBS, added with fresh media (MEM supplemented with antibiotics only) having 5 iM of AG879 or 0.1% DM9J (vehicle control) at indicated times. Supernatant was collected at 10 hpi and quantified in Vero cells by plaque assay.

In recent years, significant progress has been reported in clinical and preclinical studies of RTKIs as anticancer therapeutics. This study contributes to suggesting RTKs as a host target for the development of novel antiviral drugs against PPR. Peste des Petits Ruminants (PPR) virus causes an acute, highly contagious and fatal disease (10-100% mortality) in small ruminants that results in loss of production and in turn increases poverty in some of the poorest part of the world. Though the vaccine is commercially available but suffers from a set of drawbacks, an important one of which is the time needed to trigger the immune response (immunity-gap). Effective, rapid control tools are, therefore, urgently needed and antiviral compounds could serve this purpose.

Currently, there is no antiviral medication available against PPR to protect the affected animals and to reduce the risk of virus spread in contact with susceptible animals during an epidemic. Receptor tyrosine kinases (RTK) are a group of growth factor receptors present on the host cell surface which upon binding of a suitable ligand leads to autophosphorylation, dimerization and subsequent activation of downstream signaling pathways. Nevertheless, RTK signaling has recently been shown to play an important role during certain viral infections. Tryphostin AG879 is a small molecule cell signaling inhibitor that specifically blocks the nerve growth factor receptor (NGFR or TrkA, a family member of RTK) signaling. In this study, we found that this inhibitor does not exhibit any cytotoxicity up to 5μM in Vero cells (as determined by MTT assay) and also does not have any virucidal activity on the extracellular virus at $\geq$ 81 μM.

Using a cell culture and virus infection model, we first time observed that blocking RTK signaling by tryphostin AG879 leads to impaired PPR virus (PPRV) replication ($P<0.001$) which suggested that RTK signaling is required for efficient PPRV replication. Further, in order to confirm whether the inhibitory effect of AG879 on PPRV replication is due to action on its target RTK and not due to off-target effects, we performed a small hairpin RNA (shRNA) knockdown experiment. Transfection of RTK (TrkA) siRNA expressing plasmid (psiRNAhH1-TrkA) into Vero cells significantly knockdown the level of the expression of TrkA (mRNA level as detected by RT-PCR) and hence, on PPRV infection, significantly impaired its replication ($P<0.05$). In addition, with the help of a time-course experiment, we first time determined that the PPRV life cycle is 6-8 hrs. Subsequent time-course experiments suggested that the RTK inhibitor tryphostin AG879 inhibit multiple post-entry steps of the PPRV life cycle. Therefore we proposed the RTK as a host target for the development of novel antiviral therapeutics against PPR.

REFERENCES

1. Abd El-Rahim, I.H., Sharawi, S.S., Barakat, M.R., El-Nahas, E.M., 2010. An outbreak of peste des petits ruminants in migratory flocks of sheep and goats in Egypt in 2006. Rev Sci Tech 29, 655-662.
2. Abu Elzein, E.M., Al-Naeem, A., 2009. Utilization of protein-A in immuno-histochemical techniques for detection of Peste des Petits Ruminants (PPR) virus antigens in tissues of experimentally infected goats. Tropical animal health and production 41, 1-4.
3. Abu Elzein, E.M., Hassanien, M.M., Al-Afaleq, A.I., Abd Elhadi, M.A., Housawi, F.M., 1990. Isolation of peste des petits ruminants from goats in Saudi Arabia. The Veterinary record 127, 309-310.
4. Abubakar, M., Arshed, M.J., Zahur, A.B., Ali, Q., Banyard, A.C., 2012. Natural infection with peste des petits ruminants virus: A pre and post vaccinal assessment following an outbreak scenario. Virus research 167, 43-47.
5. Abubakar, M., Jamal, S.M., Arshed, M.J., Hussain, M., Ali, Q., 2009. Peste des petits ruminants virus (PPRV) infection; its association with species, seasonal variations and geography. Tropical animal health and production 41, 1197-1202.
6. Allen, I.C., Scull, M.A., Moore, C.B., Holl, E.K., McElvania-TeKippe, E., Taxman, D.J., Guthrie, E.H., Pickles, R.J., Ting, J.P., 2009. The NLRP3 inflammasome mediates in vivo innate immunity to influenza A virus through recognition of viral RNA. Immunity 30, 556-565.
7. Amjad, H., Qamar ul, I., Forsyth, M., Barrett, T., Rossiter, P.B., 1996. Peste des petits ruminants in goats in Pakistan. The Veterinary record 139, 118-119.
8. Anderson, J., McKay, J.A., 1994. The detection of antibodies against peste des petits ruminants virus in cattle, sheep and goats and the possible implications to rinderpest control programmes. Epidemiology and infection 112, 225-231.
9. Aoshi, T., Koyama, S., Kobiyama, K., Akira, S., Ishii, K.J., 2011. Innate and adaptive immune responses to viral infection and vaccination. Current opinion in virology 1, 226-232.
10. Ayari-Fakhfakh, E., Ghram, A., Bouattour, A., Larbi, I., Gribaa-Dridi, L., Kwiatek, O., Bouloy, M., Libeau, G., Albina, E., Cetre-Sossah, C., 2011. First serological investigation of peste-des-petits-ruminants and Rift Valley fever in Tunisia. Vet J 187, 402-404.

11. Bailey, D., Banyard, A., Dash, P., Ozkul, A., Barrett, T., 2005. Full genome sequence of peste des petits ruminants virus, a member of the Morbillivirus genus. Virus research 110, 119-124.
12. Balamurugan, V., Krishnamoorthy, P., Veeregowda, B.M., Sen, A., Rajak, K.K., Bhanuprakash, V., Gajendragad, M.R., Prabhudas, K., 2012a. Seroprevalence of Peste des petits ruminants in cattle and buffaloes from Southern Peninsular India. Tropical animal health and production 44, 301-306.
13. Balamurugan, V., Saravanan, P., Sen, A., Rajak, K.K., Bhanuprakash, V., Krishnamoorthy, P., Singh, R.K., 2011. Sero-epidemiological study of peste des petits ruminants in sheep and goats in India between 2003 and 2009. Rev Sci Tech 30, 889-896.
14. Balamurugan, V., Sen, A., Venkatesan, G., Bhanot, V., Yadav, V., Bhanuprakash, V., Singh, R.K., 2012b. Peste des petits ruminants virus detected in tissues from an Asiatic lion (Panthera leo persica) belongs to Asian lineage IV. Journal of veterinary science 13, 203-206.
15. Balamurugan, V., Sen, A., Venkatesan, G., Yadav, V., Bhanot, V., Bhanuprakash, V., Singh, R.K., 2010a. Application of semi-quantitative M gene-based hydrolysis probe (TaqMan) real-time RT-PCR assay for the detection of peste des petits ruminants virus in the clinical samples for investigation into clinical prevalence of disease. Transboundary and emerging diseases 57, 383-395.
16. Balamurugan, V., Sen, A., Venkatesan, G., Yadav, V., Bhanot, V., Bhanuprakash, V., Singh, R.K., 2012c. A rapid and sensitive one step-SYBR green based semi quantitative real time RT-PCR for the detection of peste des petits ruminants virus in the clinical samples. Virologica Sinica 27, 1-9.
17. Balamurugan, V., Sen, A., Venkatesan, G., Yadav, V., Bhanot, V., Riyesh, T., Bhanuprakash, V., Singh, R.K., 2010b. Sequence and phylogenetic analyses of the structural genes of virulent isolates and vaccine strains of peste des petits ruminants virus from India. Transboundary and emerging diseases 57, 352-364.
18. Balamurugan, V., Sen, A., Venkatesan, G., Yadav, V., Bhanuprakash, V., Singh, R.K., 2010c. Isolation and identification of virulent peste des petits ruminants viruses from PPR outbreaks in India. Tropical animal health and production 42, 1043-1046.
19. Band, C.J., Posner, B.I., Dumas, V., Contreres, J.O., 1997. Early signaling events triggered by peroxovanadium [bpV(phen)] are

insulin receptor kinase (IRK)-dependent: specificity of inhibition of IRK-associated protein tyrosine phosphatase(s) by bpV(phen). Mol Endocrinol 11, 1899-1910.

20. Bao, J., Wang, Z., Li, L., Wu, X., Sang, P., Wu, G., Ding, G., Suo, L., Liu, C., Wang, J., Zhao, W., Li, J., Qi, L., 2011a. Detection and genetic characterization of peste des petits ruminants virus in free-living bharals (*Pseudois nayaur*) in Tibet, China. Research in veterinary science 90, 238-240.
21. Bao, J.Y., Wang, Z.L., Li, L., Zhao, W.J., Suo, L.C., Li, J.M., Wang, Y.L., Wu, X.D., Liu, C.J., Liu, Y.T., Yu, X.J., Yang, Y.M., 2008. [Sequence analysis of the nucleocapsid gene and genome promoter region of peste des petits ruminants virus of Chinese origin]. Bing du xue bao = Chinese journal of virology / [bian ji, Bing du xue bao bian ji wei yuan hui] 24, 464-471.
22. Bao, J.Y., Zhao, W.J., Li, L., Wang, Z.L., Wu, G.Z., Wu, X.D., Liu, C.J., Wang, Q.H., Wang, J.W., Liu, Y.T., Li, J.M., Wang, Y.L., 2011b. [Sequence analysis of the phosphoprotein gene of peste des petits ruminants virus of Chinese origin]. Bing du xue bao = Chinese journal of virology / [bian ji, Bing du xue bao bian ji wei yuan hui] 27, 26-33.
23. Bhanuprakash, V., Saravanan, P., Hosamani, M., Balamurugan, V., Mondal, B., Singh, R.K., 2008. Status of sheep sera to bluetongue, peste des petits ruminants and sheep pox in a few northern states of India. Veterinaria italiana 44, 527-536.
24. Biarc, J., Chalkley, R.J., Burlingame, A.L., Bradshaw, R.A., 2011. Receptor tyrosine kinase signaling--a proteomic perspective. Advances in enzyme regulation 51, 293-305.
25. Bloom, J.D., Gong, L.I., Baltimore, D., 2010. Permissive secondary mutations enable the evolution of influenza oseltamivir resistance. Science 328, 1272-1275.
26. Blume-Jensen, P., Hunter, T., 2001. Oncogenic kinase signalling. Nature 411, 355-365.
27. Bodjo, S.C., Lelenta, M., Couacy-Hymann, E., Kwiatek, O., Albina, E., Gargani, D., Libeau, G., Diallo, A., 2008. Mapping the Peste des Petits Ruminants virus nucleoprotein: identification of two domains involved in protein self-association. Virus research 131, 23-32.
28. Brass, A.L., Huang, I.C., Benita, Y., John, S.P., Krishnan, M.N., Feeley, E.M., Ryan, B.J., Weyer, J.L., van der Weyden, L., Fikrig, E., Adams,

D.J., Xavier, R.J., Farzan, M., Elledge, S.J., 2009. The IFITM proteins mediate cellular resistance to influenza A H1N1 virus, West Nile virus, and dengue virus. Cell 139, 1243-1254.

29. Braunger, J., Schleithoff, L., Schulz, A.S., Kessler, H., Lammers, R., Ullrich, A., Bartram, C.R., Janssen, J.W., 1997. Intracellular signaling of the Ufo/Axl receptor tyrosine kinase is mediated mainly by a multi-substrate docking-site. Oncogene 14, 2619-2631.
30. Cam, Y., Gencay, A., Beyaz, L., Atalay, O., Atasever, A., Ozkul, A., Kibar, M., 2005. Peste des petits ruminants in a sheep and goat flock in Kayseri province, Turkey. The Veterinary record 157, 523-524.
31. Chan, P., 2010. Exotic animal diseases bulletin. Peste des Petits Ruminants. Australian veterinary journal 88, N20-21.
32. Charleston, B., Bankowski, B.M., Gubbins, S., Chase-Topping, M.E., Schley, D., Howey, R., Barnett, P.V., Gibson, D., Juleff, N.D., Woolhouse, M.E., 2011. Relationship between clinical signs and transmission of an infectious disease and the implications for control. Science 332, 726-729.
33. Coen, D.M., Whitley, R.J., 2011. Antiviral drugs and antiviral drug resistance. Current opinion in virology 1, 545-547.
34. Cohen-Daniel, L., Zakay-Rones, Z., Resnick, I.B., Shapira, M.Y., Dorozhko, M., Mador, N., Greenbaum, E., Wolf, D.G., 2009. Emergence of oseltamivir-resistant influenza A/H3N2 virus with altered hemagglutination pattern in a hematopoietic stem cell transplant recipient. Journal of clinical virology : the official publication of the Pan American Society for Clinical Virology 44, 138-140.
35. Couacy-Hymann, E., Bodjo, C., Danho, T., Libeau, G., Diallo, A., 2005. Surveillance of wildlife as a tool for monitoring rinderpest and peste des petits ruminants in West Africa. Rev Sci Tech 24, 869-877.
36. Couacy-Hymann, E., Bodjo, C., Danho, T., Libeau, G., Diallo, A., 2007a. Evaluation of the virulence of some strains of peste-des-petits-ruminants virus (PPRV) in experimentally infected West African dwarf goats. Vet J 173, 178-183.
37. Couacy-Hymann, E., Bodjo, S.C., Danho, T., Koffi, M.Y., Libeau, G., Diallo, A., 2007b. Early detection of viral excretion from experimentally infected goats with peste-des-petits ruminants virus. Preventive veterinary medicine 78, 85-88.

38. Crumpacker, C., 1983. Resistance of herpes simplex virus to antiviral agents. Is it clinically important? Drugs 26, 373-377.
39. Cuevas, J.M., Sanjuan, R., Moya, A., Elena, S.F., 2005. Mode of selection and experimental evolution of antiviral drugs resistance in vesicular stomatitis virus. Infection, genetics and evolution : journal of molecular epidemiology and evolutionary genetics in infectious diseases 5, 55-65.
40. Daikoku, T., Shiraki, K., 2007. [Antiviral resistance of herpes simplex virus and varicella-zoster virus]. Nihon rinsho. Japanese journal of clinical medicine 65 Suppl 2 Pt. 1, 480-486.
41. Darby, G., Larder, B.A., 1992. The clinical significance of antiviral drug resistance. Research in virology 143, 116-120.
42. De Nardi, M., Lamin Saleh, S.M., Batten, C., Oura, C., Di Nardo, A., Rossi, D., 2012. First evidence of peste des petits ruminants (PPR) virus circulation in Algeria (Sahrawi territories): outbreak investigation and virus lineage identification. Transboundary and emerging diseases 59, 214-222.
43. Delil, F., Asfaw, Y., Gebreegziabher, B., 2012. Prevalence of antibodies to peste des petits ruminants virus before and during outbreaks of the disease in Awash Fentale district, Afar, Ethiopia. Tropical animal health and production.
44. Demirov, D., Gabriel, G., Schneider, C., Hohenberg, H., Ludwig, S., 2012. Interaction of influenza A virus matrix protein with RACK1 is required for virus release. Cellular microbiology 14, 774-789.
45. Diallo, A., 2006. Control of peste des petits ruminants and poverty alleviation? Journal of veterinary medicine. B, Infectious diseases and veterinary public health 53 Suppl 1, 11-13.
46. Diallo, A., Minet, C., Le Goff, C., Berhe, G., Albina, E., Libeau, G., Barrett, T., 2007. The threat of peste des petits ruminants: progress in vaccine development for disease control. Vaccine 25, 5591-5597.
47. Drew, W.L., 1996. Cytomegalovirus resistance to antiviral therapies. American journal of health-system pharmacy : AJHP : official journal of the American Society of Health-System Pharmacists 53, S17-23.
48. Droebner, K., Pleschka, S., Ludwig, S., Planz, O., 2011. Antiviral activity of the MEK-inhibitor U0126 against pandemic H1N1v and highly pathogenic avian influenza virus *in vitro* and in vivo. Antiviral research 92, 195-203.

49. Ehrhardt, C., Marjuki, H., Wolff, T., Nurnberg, B., Planz, O., Pleschka, S., Ludwig, S., 2006. Bivalent role of the phosphatidylinositol-3-kinase (PI3K) during influenza virus infection and host cell defence. Cellular microbiology 8, 1336-1348.
50. Ehrhardt, C., Seyer, R., Hrincius, E.R., Eierhoff, T., Wolff, T., Ludwig, S., 2010. Interplay between influenza A virus and the innate immune signaling. Microbes and infection / Institut Pasteur 12, 81-87.
51. Ehrhardt, C., Wolff, T., Pleschka, S., Planz, O., Beermann, W., Bode, J.G., Schmolke, M., Ludwig, S., 2007. Influenza A virus NS1 protein activates the PI3K/Akt pathway to mediate antiapoptotic signaling responses. Journal of virology 81, 3058-3067.
52. Eierhoff, T., Ludwig, S., Ehrhardt, C., 2009. The influenza A virus matrix protein as a marker to monitor initial virus internalisation. Biological chemistry 390, 509-515.
53. Fischer, K.P., 1999. Resistance of hepatitis B to antiviral agents. Hepatology 29, 996.
54. Flicker, S.H., Schneider, S.M., Offterdinger, M., Dittrich, E., Fazeny, B., Valenta, R., Huber, H., Dittrich, C., Grunt, T.W., 1997. Tyrosine kinase signaling pathways control the expression of retinoic acid receptor-alpha in SK-BR-3 breast cancer cells. Cancer letters 115, 63-72.
55. Flory, E., Kunz, M., Scheller, C., Jassoy, C., Stauber, R., Rapp, U.R., Ludwig, S., 2000. Influenza virus-induced NF-kappaB-dependent gene expression is mediated by overexpression of viral proteins and involves oxidative radicals and activation of IkappaB kinase. The Journal of biological chemistry 275, 8307-8314.
56. Freifeld, A.G., Ostrove, J.M., 1991. Resistance of viruses to antiviral drugs. Annual review of medicine 42, 247-259.
57. Fry, A.M., Gubareva, L.V., 2012. Understanding influenza virus resistance to antiviral agents; early warning signs for wider community circulation. The Journal of infectious diseases.
58. Furley, C.W., Taylor, W.P., Obi, T.U., 1987. An outbreak of peste des petits ruminants in a zoological collection. The Veterinary record 121, 443-447.
59. Gibbs, E.P., Taylor, W.P., Lawman, M.J., Bryant, J., 1979. Classification of peste des petits ruminants virus as the fourth member of the genus Morbillivirus. Intervirology 11, 268-274.

60. Goris, N., Vandenbussche, F., De Clercq, K., 2008. Potential of antiviral therapy and prophylaxis for controlling RNA viral infections of livestock. Antiviral research 78, 170-178.
61. Griffiths, P.D., 2009. A perspective on antiviral resistance. Journal of clinical virology : the official publication of the Pan American Society for Clinical Virology 46, 3-8.
62. Grimes, M.L., Miettinen, H.M., 2003. Receptor tyrosine kinase and G-protein coupled receptor signaling and sorting within endosomes. Journal of neurochemistry 84, 905-918.
63. Guillot, L., Le Goffic, R., Bloch, S., Escriou, N., Akira, S., Chignard, M., Si-Tahar, M., 2005. Involvement of toll-like receptor 3 in the immune response of lung epithelial cells to double-stranded RNA and influenza A virus. The Journal of biological chemistry 280, 5571-5580.
64. Hader, T., Wainwright, D., Shandala, T., Saint, R., Taubert, H., Bronner, G., Jackle, H., 2000. Receptor tyrosine kinase signaling regulates different modes of Groucho-dependent control of Dorsal. Current biology : CB 10, 51-54.
65. Hamelin, M.E., Baz, M., Abed, Y., Couture, C., Joubert, P., Beaulieu, E., Bellerose, N., Plante, M., Mallett, C., Schumer, G., Kobinger, G.P., Boivin, G., 2010. Oseltamivir-resistant pandemic A/H1N1 virus is as virulent as its wild-type counterpart in mice and ferrets. PLoS pathogens 6, e1001015.
66. Hammouchi, M., Loutfi, C., Sebbar, G., Touil, N., Chaffai, N., Batten, C., Harif, B., Oura, C., El Harrak, M., 2012. Experimental infection of alpine goats with a Moroccan strain of peste des petits ruminants virus (PPRV). Veterinary microbiology.
67. Hao, L., Sakurai, A., Watanabe, T., Sorensen, E., Nidom, C.A., Newton, M.A., Ahlquist, P., Kawaoka, Y., 2008. Drosophila RNAi screen identifies host genes important for influenza virus replication. Nature 454, 890-893.
68. Hayakawa, M., Miyashita, H., Sakamoto, I., Kitagawa, M., Tanaka, H., Yasuda, H., Karin, M., Kikugawa, K., 2003. Evidence that reactive oxygen species do not mediate NF-kappaB activation. The EMBO journal 22, 3356-3366.
69. Hayden, F., 2009. Developing new antiviral agents for influenza treatment: what does the future hold? Clinical infectious diseases : an official publication of the Infectious Diseases Society of America 48 Suppl 1, S3-13.

70. Hayden, F.G., 2006. Antivirals for influenza: historical perspectives and lessons learned. Antiviral research 71, 372-378.
71. Hayden, F.G., de Jong, M.D., 2011. Emerging influenza antiviral resistance threats. The Journal of infectious diseases 203, 6-10.
72. Hayden, F.G., Hay, A.J., 1992. Emergence and transmission of influenza A viruses resistant to amantadine and rimantadine. Current topics in microbiology and immunology 176, 119-130.
73. Hegde, R., Gomes, A.R., Muniyellappa, H.K., Byregowda, S.M., Giridhar, P., Renukaprasad, C., 2009. A short note on peste des petits ruminants in Karnataka, India. Rev Sci Tech 28, 1031-1035.
74. Herold, S., Ludwig, S., Pleschka, S., Wolff, T., 2012. Apoptosis signaling in influenza virus propagation, innate host defense, and lung injury. Journal of leukocyte biology.
75. Hirsch, M.S., Schooley, R.T., 1989. Resistance to antiviral drugs: the end of innocence. The New England journal of medicine 320, 313-314.
76. Hoffmann, B., Wiesner, H., Maltzan, J., Mustefa, R., Eschbaumer, M., Arif, F.A., Beer, M., 2012. Fatalities in wild goats in Kurdistan associated with Peste des Petits Ruminants virus. Transboundary and emerging diseases 59, 173-176.
77. Hopfner, M., Sutter, A.P., Huether, A., Baradari, V., Scherubl, H., 2006. Tyrosine kinase of insulin-like growth factor receptor as target for novel treatment and prevention strategies of colorectal cancer. World journal of gastroenterology : WJG 12, 5635-5643.
78. Hornung, V., Ellegast, J., Kim, S., Brzozka, K., Jung, A., Kato, H., Poeck, H., Akira, S., Conzelmann, K.K., Schlee, M., Endres, S., Hartmann, G., 2006. 5'-Triphosphate RNA is the ligand for RIG-I. Science 314, 994-997.
79. Hovi, T., Jarvinen, A., Pyhala, R., Ristola, M., Salminen, M., 2002. [Viruses and antiviral drug resistance]. Duodecim; laaketieteellinen aikakauskirja 118, 911-918.
80. Hrincius, E.R., Dierkes, R., Anhlan, D., Wixler, V., Ludwig, S., Ehrhardt, C., 2011. Phosphatidylinositol-3-kinase (PI3K) is activated by influenza virus vRNA via the pathogen pattern receptor Rig-I to promote efficient type I interferon production. Cellular microbiology 13, 1907-1919.
81. Ichinohe, T., Ainai, A., Nakamura, T., Akiyama, Y., Maeyama, J., Odagiri, T., Tashiro, M., Takahashi, H., Sawa, H., Tamura, S., Chiba, J.,

Kurata, T., Sata, T., Hasegawa, H., 2010. Induction of cross-protective immunity against influenza A virus H5N1 by an intranasal vaccine with extracts of mushroom mycelia. Journal of medical virology 82, 128-137.

82. Ireton, R.C., Chen, J., 2005. EphA2 receptor tyrosine kinase as a promising target for cancer therapeutics. Current cancer drug targets 5, 149-157.

83. Ismail, A.M., Samuel, P., Eapen, C.E., Kannangai, R., Abraham, P., 2012. Antiviral resistance mutations and genotype-associated amino acid substitutions in treatment-naive hepatitis B virus-infected individuals from the Indian subcontinent. Intervirology 55, 36-44.

84. Jagtap, S.P., Rajak, K.K., Garg, U.K., Sen, A., Bhanuprakash, V., Sudhakar, S.B., Balamurugan, V., Patel, A., Ahuja, A., Singh, R.K., Vanamayya, P.R., 2012. Effect of immunosuppression on pathogenesis of peste des petits ruminants (PPR) virus infection in goats. Microbial pathogenesis 52, 217-226.

85. Jensen, S., Thomsen, A.R., 2012. Sensing of RNA viruses: a review of innate immune receptors involved in recognizing RNA virus invasion. Journal of virology 86, 2900-2910.

86. Kamakura, S., Moriguchi, T., Nishida, E., 1999. Activation of the protein kinase ERK5/BMK1 by receptor tyrosine kinases. Identification and characterization of a signaling pathway to the nucleus. The Journal of biological chemistry 274, 26563-26571.

87. Karlas, A., Machuy, N., Shin, Y., Pleissner, K.P., Artarini, A., Heuer, D., Becker, D., Khalil, H., Ogilvie, L.A., Hess, S., Maurer, A.P., Muller, E., Wolff, T., Rudel, T., Meyer, T.F., 2010. Genome-wide RNAi screen identifies human host factors crucial for influenza virus replication. Nature 463, 818-822.

88. Kawai, Y., Minami, T., Fujimori, M., Hosaka, K., Mizuno, R., Ikomi, F., Kodama, T., Ohhashi, T., 2007. Characterization and microarray analysis of genes in human lymphatic endothelial cells from patients with breast cancer. Lymphatic research and biology 5, 115-126.

89. Khalafalla, A.I., Saeed, I.K., Ali, Y.H., Abdurrahman, M.B., Kwiatek, O., Libeau, G., Obeida, A.A., Abbas, Z., 2010. An outbreak of peste des petits ruminants (PPR) in camels in the Sudan. Acta tropica 116, 161-165.

90. Khan, H.A., Siddique, M., Sajjad ur, R., Abubakar, M., Ashraf, M., 2008. The detection of antibody against peste des petits ruminants virus in sheep, goats, cattle and buffaloes. Tropical animal health and production 40, 521-527.

91. Kim, Y.C., Park, K.O., Bom, H.S., Lim, S.C., Park, H.K., Na, H.J., Park, J.H., 1997. Combining ADA, protein and IFN-gamma best allows discrimination between tuberculous and malignant pleural effusion. The Korean journal of internal medicine 12, 225-231.

92. Kimberlin, D.W., Coen, D.M., Biron, K.K., Cohen, J.I., Lamb, R.A., McKinlay, M., Emini, E.A., Whitley, R.J., 1995a. Molecular mechanisms of antiviral resistance. Antiviral research 26, 369-401.

93. Kimberlin, D.W., Couch, R.B., Galasso, G.J., Richman, D.D., Whitley, R.J., 1995b. Research initiatives in studies of antiviral resistance and consensus points and recommendations. Antiviral research 26, 439-452.

94. Kimberlin, D.W., Kern, E.R., Sidwell, R.W., North, T.W., Whitley, R.J., 1995c. Models of antiviral resistance. Antiviral research 26, 415-422.

95. Kimberlin, D.W., Whitley, R.J., 1995. Antiviral resistance--an emerging problem. Antiviral research 26, 365-368.

96. Kimberlin, D.W., Whitley, R.J., 1996. Antiviral resistance: mechanisms, clinical significance, and future implications. The Journal of antimicrobial chemotherapy 37, 403-421.

97. Kinne, J., Kreutzer, R., Kreutzer, M., Wernery, U., Wohlsein, P., 2010. Peste des petits ruminants in Arabian wildlife. Epidemiology and infection 138, 1211-1214.

98. Konig, R., Stertz, S., Zhou, Y., Inoue, A., Hoffmann, H.H., Bhattacharyya, S., Alamares, J.G., Tscherne, D.M., Ortigoza, M.B., Liang, Y., Gao, Q., Andrews, S.E., Bandyopadhyay, S., De Jesus, P., Tu, B.P., Pache, L., Shih, C., Orth, A., Bonamy, G., Miraglia, L., Ideker, T., Garcia-Sastre, A., Young, J.A., Palese, P., Shaw, M.L., Chanda, S.K., 2010. Human host factors required for influenza virus replication. Nature 463, 813-817.

99. Koyama, S., Ishii, K.J., Coban, C., Akira, S., 2008. Innate immune response to viral infection. Cytokine 43, 336-341.

100. Krystal, G.W., 2005. Receptor tyrosine kinase mutations: molecular indicators for therapeutic response or signposts marking important signaling pathways? Cancer biology & therapy 4, 1417-1418.

101. Kuida, K., Lippke, J.A., Ku, G., Harding, M.W., Livingston, D.J., Su, M.S., Flavell, R.A., 1995. Altered cytokine export and apoptosis in mice deficient in interleukin-1 beta converting enzyme. Science 267, 2000-2003.
102. Kulkarni, D.D., Bhikane, A.U., Shaila, M.S., Varalakshmi, P., Apte, M.P., Narladkar, B.W., 1996. Peste des petits ruminants in goats in India. The Veterinary record 138, 187-188.
103. Kumar, N., Liang, Y., Parslow, T.G., 2011a. Receptor tyrosine kinase inhibitors block multiple steps of influenza a virus replication. Journal of virology 85, 2818-2827.
104. Kumar, N., Sharma, N.R., Ly, H., Parslow, T.G., Liang, Y., 2011b. Receptor tyrosine kinase inhibitors that block replication of influenza a and other viruses. Antimicrobial agents and chemotherapy 55, 5553-5559.
105. Kumar, N., Xin, Z.T., Liang, Y., Ly, H., 2008. NF-kappaB signaling differentially regulates influenza virus RNA synthesis. Journal of virology 82, 9880-9889.
106. Kwiatek, O., Minet, C., Grillet, C., Hurard, C., Carlsson, E., Karimov, B., Albina, E., Diallo, A., Libeau, G., 2007. Peste des petits ruminants (PPR) outbreak in Tajikistan. Journal of comparative pathology 136, 111-119.
107. Lefevre, P.C., Diallo, A., 1990. Peste des petits ruminants. Rev Sci Tech 9, 935-981.
108. Lemmon, M.A., Schlessinger, J., 2010. Cell signaling by receptor tyrosine kinases. Cell 141, 1117-1134.
109. Li, H., Willingham, S.B., Ting, J.P., Re, F., 2008. Cutting edge: inflammasome activation by alum and alum's adjuvant effect are mediated by NLRP3. J Immunol 181, 17-21.
110. Li, W.X., 2005. Functions and mechanisms of receptor tyrosine kinase Torso signaling: lessons from Drosophila embryonic terminal development. Developmental dynamics : an official publication of the American Association of Anatomists 232, 656-672.
111. Liu, F.X., Liu, Z.S., Wang, Z.L., 2012. [Feasibility study on global peste des petits ruminants eradication based on rinderpest eradication]. Bing du xue bao = Chinese journal of virology / [bian ji, Bing du xue bao bian ji wei yuan hui] 28, 89-96.

112. Locarnini, S., Bowden, S., 2010. Drug resistance in antiviral therapy. Clinics in liver disease 14, 439-459.

113. Lu, X., Masic, A., Li, Y., Shin, Y., Liu, Q., Zhou, Y., 2010. The PI3K/Akt pathway inhibits influenza A virus-induced Bax-mediated apoptosis by negatively regulating the JNK pathway via ASK1. The Journal of general virology 91, 1439-1449.

114. Ludwig, S., 2011. Disruption of virus-host cell interactions and cell signaling pathways as an anti-viral approach against influenza virus infections. Biological chemistry 392, 837-847.

115. Ludwig, S., Ehrhardt, C., Neumeier, E.R., Kracht, M., Rapp, U.R., Pleschka, S., 2001a. Influenza virus-induced AP-1-dependent gene expression requires activation of the JNK signaling pathway. The Journal of biological chemistry 276, 10990-10998.

116. Ludwig, S., Ehrhardt, C., Neumeier, E.R., Kracht, M., Rapp, U.R., Pleschka, S., 2001b. Influenza virus-induced AP-1-dependent gene expression requires activation of the JNK signaling pathway. The Journal of biological chemistry 276, 10990-10998.

117. Ludwig, S., Planz, O., Pleschka, S., Wolff, T., 2003. Influenza-virus-induced signaling cascades: targets for antiviral therapy? Trends in molecular medicine 9, 46-52.

118. Ludwig, S., Pleschka, S., Planz, O., Wolff, T., 2006. Ringing the alarm bells: signalling and apoptosis in influenza virus infected cells. Cellular microbiology 8, 375-386.

119. Ludwig, S., Pleschka, S., Wolff, T., 1999. A fatal relationship--influenza virus interactions with the host cell. Viral immunology 12, 175-196.

120. Ludwig, S., Wolff, T., Ehrhardt, C., Wurzer, W.J., Reinhardt, J., Planz, O., Pleschka, S., 2004. MEK inhibition impairs influenza B virus propagation without emergence of resistant variants. FEBS letters 561, 37-43.

121. Luig, C., Kother, K., Dudek, S.E., Gaestel, M., Hiscott, J., Wixler, V., Ludwig, S., 2010. MAP kinase-activated protein kinases 2 and 3 are required for influenza A virus propagation and act via inhibition of PKR. FASEB journal : official publication of the Federation of American Societies for Experimental Biology 24, 4068-4077.

122. Luka, P.D., Erume, J., Mwiine, F.N., Ayebazibwe, C., 2012. Molecular characterization of peste des petits ruminants virus from the Karamoja region of Uganda (2007-2008). Archives of virology 157, 29-35.

123. Mackiewicz, M., Huppi, K., Pitt, J.J., Dorsey, T.H., Ambs, S., Caplen, N.J., 2011. Identification of the receptor tyrosine kinase AXL in breast cancer as a target for the human miR-34a microRNA. Breast cancer research and treatment 130, 663-679.
124. Mahapatra, M., Parida, S., Baron, M.D., Barrett, T., 2006. Matrix protein and glycoproteins F and H of Peste-des-petits-ruminants virus function better as a homologous complex. The Journal of general virology 87, 2021-2029.
125. Malik, Y.S., Singh, D., Chandrashekar, K.M., Shukla, S., Sharma, K., Vaid, N., Chakravarti, S., 2011. Occurrence of dual infection of peste-des-petits-ruminants and goatpox in indigenous goats of central India. Transboundary and emerging diseases 58, 268-273.
126. Marjuki, H., Alam, M.I., Ehrhardt, C., Wagner, R., Planz, O., Klenk, H.D., Ludwig, S., Pleschka, S., 2006. Membrane accumulation of influenza A virus hemagglutinin triggers nuclear export of the viral genome via protein kinase Calpha-mediated activation of ERK signaling. The Journal of biological chemistry 281, 16707-16715.
127. Marjuki, H., Yen, H.L., Franks, J., Webster, R.G., Pleschka, S., Hoffmann, E., 2007. Higher polymerase activity of a human influenza virus enhances activation of the hemagglutinin-induced Raf/MEK/ERK signal cascade. Virology Journal 4, 134.
128. Martinon, F., Mayor, A., Tschopp, J., 2009. The inflammasomes: guardians of the body. Annual review of immunology 27, 229-265.
129. Mayer, B.J., 2012. Perspective: Dynamics of receptor tyrosine kinase signaling complexes. FEBS letters.
130. McCall, S.H., Sahraei, M., Young, A.B., Worley, C.S., Duncan, J.A., Ting, J.P., Marriott, I., 2008. Osteoblasts express NLRP3, a nucleotide-binding domain and leucine-rich repeat region containing receptor implicated in bacterially induced cell death. Journal of bone and mineral research: the official journal of the American Society for Bone and Mineral Research 23, 30-40.
131. Mitzner, D., Dudek, S.E., Studtrucker, N., Anhlan, D., Mazur, I., Wissing, J., Jansch, L., Wixler, L., Bruns, K., Sharma, A., Wray, V., Henklein, P., Ludwig, S., Schubert, U., 2009. Phosphorylation of the influenza A virus protein PB1-F2 by PKC is crucial for apoptosis promoting functions in monocytes. Cellular microbiology 11, 1502-1516.

132. Mizumura, K., Hashimoto, S., Maruoka, S., Gon, Y., Kitamura, N., Matsumoto, K., Hayashi, S., Shimizu, K., Horie, T., 2003. Role of mitogen-activated protein kinases in influenza virus induction of prostaglandin E2 from arachidonic acid in bronchial epithelial cells. Clinical and experimental allergy : journal of the British Society for Allergy and Clinical Immunology 33, 1244-1251.

133. Mogensen, P.H., Jakobsen, J., 2001. Cued and non-cued repetitive ballistic movements. A kinematic study in healthy subjects. Acta neurologica Scandinavica 103, 12-19.

134. Munir, M., Zohari, S., Suluku, R., Leblanc, N., Kanu, S., Sankoh, F.A., Berg, M., Barrie, M.L.,Stahl, K., 2012. Genetic characterization of peste des petits ruminants virus, Sierra Leone. Emerging infectious diseases 18, 193-195.

135. Munk, C., Sommer, A.F., Konig, R., 2011. Systems-biology approaches to discover anti-viral effectors of the human innate immune response. Viruses 3, 1112-1130.

136. Nakachi, I., Naoki, K., Soejima, K., Kawada, I., Watanabe, H., Yasuda, H., Nakayama, S., Yoda, S., Satomi, R., Ikemura, S., Terai, H., Sato, T., Ishizaka, A., 2010. The combination of multiple receptor tyrosine kinase inhibitor and mammalian target of rapamycin inhibitor overcomes erlotinib resistance in lung cancer cell lines through c-Met inhibition. Molecular cancer research : MCR 8, 1142-1151.

137. Nanda, Y.P., Chatterjee, A., Purohit, A.K., Diallo, A., Innui, K., Sharma, R.N., Libeau, G., Thevasagayam, J.A., Bruning, A., Kitching, R.P., Anderson, J., Barrett, T., Taylor, W.P., 1996. The isolation of peste des petits ruminants virus from northern India. Veterinary microbiology 51, 207-216.

138. Nimmerjahn, F., Dudziak, D., Dirmeier, U., Hobom, G., Riedel, A., Schlee, M., Staudt, L.M., Rosenwald, A., Behrends, U., Bornkamm, G.W., Mautner, J., 2004. Active NF-kappaB signalling is a prerequisite for influenza virus infection. The Journal of general virology 85, 2347-2356.

139. Pauli, E.K., Schmolke, M., Wolff, T., Viemann, D., Roth, J., Bode, J.G., Ludwig, S., 2008. Influenza A virus inhibits type I IFN signaling via NF-kappaB-dependent induction of SOCS-3 expression. PLoS pathogens 4, e1000196.

140. Pawlotsky, J.M., 2012. Is hepatitis virus resistance to antiviral drugs a threat? Gastroenterology 142, 1369-1372.

141. Pawson, T., 1995a. Protein-tyrosine kinases. Getting down to specifics. Nature 373, 477-478.
142. Pawson, T., 1995b. Protein modules and signalling networks. Nature 373, 573-580.
143. Pichlmair, A., Schulz, O., Tan, C.P., Naslund, T.I., Liljestrom, P., Weber, F., Reis e Sousa, C., 2006. RIG-I-mediated antiviral responses to single-stranded RNA bearing 5'-phosphates. Science 314, 997-1001.
144. Pillay, D., Schinazi, R.F., 1998. Antiviral drug resistance: from the laboratory to the patient. Antiviral therapy 3, 237.
145. Planz, O., Pleschka, S., Ludwig, S., 2001. MEK-specific inhibitor U0126 blocks spread of Borna disease virus in cultured cells. Journal of virology 75, 4871-4877.
146. Pleschka, S., Wolff, T., Ehrhardt, C., Hobom, G., Planz, O., Rapp, U.R., Ludwig, S., 2001. Influenza virus propagation is impaired by inhibition of the Raf/MEK/ERK signalling cascade. Nature cell biology 3, 301-305.
147. Prober, C.G., 2004. New antivirals and antiviral resistance. Advances in experimental medicine and biology 549, 9-12.
148. Provencher, V.M., Coccaro, E., Lacasse, J.J., Schang, L.M., 2004. Antiviral drugs that target cellular proteins may play major roles in combating HIV resistance. Current pharmaceutical design 10, 4081-4101.
149. Raghavendra, A.G., Gajendragad, M.R., Sengupta, P.P., Patil, S.S., Tiwari, C.B., Balumahendiran, M., Sankri, V., Prabhudas, K., 2008. Seroepidemiology of peste des petits ruminants in sheep and goats of southern peninsular India. Rev Sci Tech 27, 861-867.
150. Raj, G.D., Kumar, A.S., Shaila, M.S., Nachimuthu, K., Palaniswami, K.S., 2003. Molecular epidemiology of peste des petits ruminants viruses from southern India. The Veterinary record 152, 264-266.
151. Raj, G.D., Rajanathan, T.M., Kumar, C.S., Ramathilagam, G., Hiremath, G., Shaila, M.S., 2008. Detection of peste des petits ruminants virus antigen using immunofiltration and antigen-competition ELISA methods. Veterinary microbiology 129, 246-251.
152. Richman, D.D., 2006. Antiviral drug resistance. Antiviral research 71, 117-121.

153. Roeder, P.L., Abraham, G., Kenfe, G., Barrett, T., 1994. Peste des petits ruminants in Ethiopian goats. Tropical animal health and production 26, 69-73.
154. Saravanan, P., Balamurugan, V., Sen, A., Sarkar, J., Sahay, B., Rajak, K.K., Hosamani, M., Yadav, M.P., Singh, R.K., 2007. Mixed infection of peste des petits ruminants and orf on a goat farm in Shahjahanpur, India. The Veterinary record 160, 410-412.
155. Schmolke, M., Garcia-Sastre, A., 2010. Evasion of innate and adaptive immune responses by influenza A virus. Cellular microbiology 12, 873-880.
156. Schmolke, M., Viemann, D., Roth, J., Ludwig, S., 2009. Essential impact of NF-kappaB signaling on the H5N1 influenza A virus-induced transcriptome. J Immunol 183, 5180-5189.
157. See, H., Wark, P., 2008. Innate immune response to viral infection of the lungs. Paediatric respiratory reviews 9, 243-250.
158. Sen, A., Saravanan, P., Balamurugan, V., Rajak, K.K., Sudhakar, S.B., Bhanuprakash, V., Parida, S., Singh, R.K., 2010. Vaccines against peste des petits ruminants virus. Expert review of vaccines 9, 785-796.
159. Shaila, M.S., Purushothaman, V., Bhavasar, D., Venugopal, K., Venkatesan, R.A., 1989. Peste des petits ruminants of sheep in India. The Veterinary record 125, 602.
160. Shaila, M.S., Shamaki, D., Forsyth, M.A., Diallo, A., Goatley, L., Kitching, R.P., Barrett, T., 1996. Geographic distribution and epidemiology of peste des petits ruminants virus. Virus research 43, 149-153.
161. Shin, Y.K., Li, Y., Liu, Q., Anderson, D.H., Babiuk, L.A., Zhou, Y., 2007. SH3 binding motif 1 in influenza A virus NS1 protein is essential for PI3K/Akt signaling pathway activation. Journal of virology 81, 12730-12739.
162. Singh, R.K., Balamurugan, V., Bhanuprakash, V., Sen, A., Saravanan, P., Pal Yadav, M., 2009. Possible control and eradication of peste des petits ruminants from India: technical aspects. Veterinaria italiana 45, 449-462.
163. Singh, R.P., 2011. Control strategies for peste des petits ruminants in small ruminants of India. Rev Sci Tech 30, 879-887.
164. Singh, R.P., Saravanan, P., Sreenivasa, B.P., Singh, R.K., Bandyopadhyay, S.K., 2004. Prevalence and distribution of peste des

petits ruminants virus infection in small ruminants in India. Rev Sci Tech 23, 807-819.

165. Strasfeld, L., Chou, S., 2010. Antiviral drug resistance: mechanisms and clinical implications. Infectious disease clinics of North America 24, 413-437.
166. Taylor, W.P., Diallo, A., Gopalakrishna, S., Sreeramalu, P., Wilsmore, A.J., Nanda, Y.P., Libeau, G., Rajasekhar, M., Mukhopadhyay, A.K., 2002. Peste des petits ruminants has been widely present in southern India since, if not before, the late 1980s. Preventive veterinary medicine 52, 305-312.
167. Ujike, M., Shimabukuro, K., Mochizuki, K., Obuchi, M., Kageyama, T., Shirakura, M., Kishida, N., Yamashita, K., Horikawa, H., Kato, Y., Fujita, N., Tashiro, M., Odagiri, T., 2010. Oseltamivir-resistant influenza viruses A (H1N1) during 2007-2009 influenza seasons, Japan. Emerging infectious diseases 16, 926-935.
168. Ullrich, A., Schlessinger, J., 1990. Signal transduction by receptors with tyrosine kinase activity. Cell 61, 203-212.
169. Wang, Z., Bao, J., Wu, X., Liu, Y., Li, L., Liu, C., Suo, L., Xie, Z., Zhao, W., Zhang, W., Yang, N., Li, J., Wang, S., Wang, J., 2009. Peste des petits ruminants virus in Tibet, China. Emerging infectious diseases 15, 299-301.
170. Watanabe, T., Watanabe, S., Kawaoka, Y., 2010. Cellular networks involved in the influenza virus life cycle. Cell host & microbe 7, 427-439.
171. Wurzer, W.J., Ehrhardt, C., Pleschka, S., Berberich-Siebelt, F., Wolff, T., Walczak, H., Planz, O., Ludwig, S., 2004. NF-kappaB-dependent induction of tumor necrosis factor-related apoptosis-inducing ligand (TRAIL) and Fas/FasL is crucial for efficient influenza virus propagation. The Journal of biological chemistry 279, 30931-30937.
172. Zahur, A.B., Ullah, A., Hussain, M., Irshad, H., Hameed, A., Jahangir, M., Farooq, M.S., 2011. Sero-epidemiology of peste des petits ruminants (PPR) in Pakistan. Preventive veterinary medicine 102, 87-92.
173. Zhang, Y., Hinojosa, M.E., Yoo, N., Holtzman, M.J., 2010. Viral and host strategies to take advantage of the innate immune response. American journal of respiratory cell and molecular biology 43, 507-510.
174. Zwick, E., Bange, J., Ullrich, A., 2001. Receptor tyrosine kinase signalling as a target for cancer intervention strategies. Endocrine-related cancer 8, 161-173.

CHAPTER 4

SEVERE ACUTE RESPIRATORY SYNDROME CORONAVIRUS-2

The coronavirus is a contagious disease caused by severe acute respiratory syndrome coronavirus-2 (SARS-CoV2), has reached pandemic status. As it spreads across the world, it has overwhelmed healthcare systems, global economic loss and led to a devastating loss of life. Coronaviridae comprises a family of enveloped, single-stranded, positive sense, RNA viruses. Coronaviruses (CoV) are canonically divided into alpha- and beta-, gamma- and delta- genera predicated on genetic clustering. The alpha- and beta-CoV are known to cause human diseases, such as common respiratory infections. The SARS-CoV2 are beta-CoV. CoVs are so- named because of the characteristic crown, or corona.

The CoV genome is organized into two parts. Highly conserved with the CoV family, the 5′ terminal end, encodes the replicase - the nonstructural proteins responsible for viral replication within the cell. It is translated as one peptide (~790 kDa) from which the constituent functional proteins are subsequently cleaved. CoV genomes encode 16 nonstructural proteins, as in SARS-CoV2, and they exhibit a multitude of functions required for viral replication.

Critical proteins for viral replication include the main protease (nsp5), the papain-like protease (nsp3) and the RNA-dependent RNA polymerase (nsp12, RdRp). The other replicase constituent proteins repurpose the cellular machinery to facilitate viral replication and to blunt the intrinsic host immune functions. The structural proteins are the constituent proteins of the transmissible viral particle or virion. The key structural CoV proteins are the nucleocapsid protein (N) and three transmembrane proteins: the spike protein (S), the membrane protein (M), and the envelope protein (E).

The S protein is responsible for virus-cell receptor interactions. The E and M proteins are responsible for membrane structure and fusion. The N protein binds viral RNA and mediates its interaction with the S, E, and M proteins for genome encapsulation.

- SARS-CoV2 is known to bind to cells, the membrane-bound glycoprotein Angiotensin Converting Enzyme 2 (ACE2).
- It has not been observed to bind other CoV receptors, namely dipeptidyl peptidase 4 (DPP4) or aminopeptidase N (APN).
- After binding of ACE2, the virus is internalized via endocytosis without access to the host intracellular compartment until a membrane fusion event occurs.
- This process is mediated, at least in part, by another membrane bound protease known as transmembrane serine protease 2 (TMPRSS2), which cleaves the S protein as a necessary step of membrane fusion.
- Interestingly, the protease activity of the CoV receptors, ACE2, DPP4 and APN, does not seem necessary for membrane fusion (14). Upon membrane fusion, the viral RNA genome enters the intracellular compartment.
- At this point, the viral RNA may be translated into its encoded structural and nonstructural proteins. The translation of the nonstructural proteins, or replicase, results in the production of a single massive polypeptide chain, from which the sixteen constituent nonstructural proteins are cleaved.
- This process is initially mediated by intracellular proteases, and then further propagated by the function of the CoV main protease and papain-like protease.
- Another replicase protein, the RNA-dependent RNA polymerase (RdRp) is responsible for the replication and amplification of the viral genome.
- During this process, mutations may be acquired by errors in replication and recombination events.

- Upon amplification of the viral RNA, more viral structural and nonstructural proteins may be generated.
- Viral structural proteins, because of their transmembrane nature (with the exception of the N protein), are targeted to the ER membrane with appropriate signal sequences.
- Viral RNA, bound by N protein, interacts with the structural proteins on the membrane of the ER and Golgi apparatus before another membrane fusion event on these membranes results in viral budding and exocytosis.

TRANSMISSION

Exposure to the Huanan seafood market was common among the earliest cases contributing to the SARS-CoV2 epidemic in China suggesting that this was a zoonotic disease with an intermediate animal host (non-aquatic animals were sold in the market). Genomic analyses have identified approximately 87% DNA sequence homology between SARS-CoV2 and two SARS-like CoV isolated from Chinese horseshoe bats, bat-SL-CoVZC45 and bat-SL-CoVZXC21, in the Zhejiang province in China. Notably, no bats are sold in the market and at the onset of the outbreak in December most bat species in Wuhan would be hibernating. Thus, similar to SARS-CoV1 and MERS-CoV, while bats may be the natural reservoir for SARS-CoV2, there is likely an unidentified intermediate animal host responsible for animal-to-human transmission.

Despite the closure of the Huanan market on January 1, 2020, the epidemic continued to expand and case clusters with no exposure to the market were reported indicating the occurrence of human-to-human transmission.

- SARS-CoV2 spreads primarily through small respiratory droplets that are expelled from infected individuals and can travel approximately 3-6 feet.
- The virus can exist in nature on surfaces and can last for up to 4 hours on copper, 24 hours on cardboard and up to 72 hours on plastic and stainless steel surfaces leading to fomite transmission.
- In fact, the Japanese National Institute of Infectious Disease reported the detection of SARS-CoV2 RNA on surfaces in the cabins of symptomatic and asymptomatic passengers on the Diamond Princess up to 17 days after they were vacated.

- A live virus has also been isolated and cultured from fecal specimens raising the possibility of oro-fecal transmission, though corroborating clinical evidence for this method of transmission is lacking.
- Airborne transmission may be facilitated in healthcare settings where aerosol-producing interventions are being performed including endotracheal intubation, bronchoscopy, suctioning, nebulizer treatment, non-invasive positive pressure ventilation and delivery of oxygen via a high-flow nasal cannula.
- These transmission data support the clinical recommendations that airborne -precautions, including the use of N-95 masks, should be implemented in these aerosol-producing settings whereas standard droplet precautions should be used during all other encounters with infected individuals.
- In a fully susceptible population, reflected by the early stages of the epidemic in China, studies have estimated a basic reproductive number (Ro) of 2.38 for SARS-CoV2, meaning that every infected individual is likely on average to spread the virus to 2 to 3 other individuals.
- An outbreak will continue to increase in size if the Ro>1. For context, seasonal influenza has a Ro of 1.5.
- Substantial transmission from asymptomatic hosts has facilitated the widespread transmission of SARS-CoV2 and contributed to its pandemic potential.
- A study from Singapore with extensive contact tracing identified 7 clusters of cases where the secondary spread of the infection occurred 1-3 days prior to symptom development in the source patient.
- Thus, containment measures aimed solely at isolating symptomatic individuals are inadequate. Furthermore, contact tracing efforts should take into account the pre-symptomatic contagious period to comprehensively capture all potentially exposed individuals.
- Ro is not a static measure and interventions including self-quarantine, contact isolation, social distancing and enhanced hygiene measures have proven to be effective in China.

Following the implementation of such measures in China, the Ro steadily decreased from 2.38 prior to January 23rd to 0.99 during the period of January 24-February 8, 2020.

Epidemiology and Clinical Manifestations of SARS-CoV2:

- The SARS-CoV2 virus has evolved rapidly since it first appeared in Wuhan, China in December 2019. What began as a few case reports of atypical pneumonia now spans the globe as a pandemic.
- Chinese Centers for Disease Control and Prevention (CDC) reported high-level details for patient characteristics, the severity of manifestations and survival in 72,314 cases of putative (47%) and confirmed (63%) COVID-19.
- In this population, predominantly identified by the presence of symptoms (~99%), <2% of cases occurred in children < 19 years of age suggesting that children either are resistant to infection or rarely symptomatic.
- Of confirmed cases, most (87%) were mild, defined by no or mild pneumonia, 14% were severe with significant infiltrates or signs of respiratory compromise and 5% were critical with respiratory failure (e.g. mechanical ventilation), shock or multiorgan system failure.
- The first confirmed case of COVID-19 in the US was identified on January 20, 2020, and the US has now surpassed all other countries in the absolute number of cases. In an early snapshot from the US CDC in 4,226 confirmed cases with symptoms or exposure, only 5% occurred in those under the age of 20.

Clinical Syndrome

Patients (1,099) from China are hospitalized multi-center case series and the most common symptoms were fever in up to 90%, followed by cough, fatigue, sputum production, and shortness of breath. Less common symptoms included headache, myalgias, sore throat, nausea, vomiting, and diarrhea. The American Association of Otolaryngology has recently highlighted anosmia and dysgeusia as possible symptoms of the disease as well.

- The median incubation period, or time from probable exposure to the first symptom, was 4 days.

- Another report detailed that 99% of infected patients develop symptoms within 14 days.
- Common lab derangements on admission included lymphopenia, elevations in C-reactive protein, lactate dehydrogenase, liver transaminases and d-dimer.
- Notably, procalcitonin was rarely elevated.
- These data are generally consistent across multiple smaller studies, several of which noted elevations in other inflammatory markers, such as IL-6, ferritin and ESR.
- Evidence of cardiac or kidney injury at admission was variable across studies but tended to be absent upon hospitalization.
- Chest computed tomography at the time of admission was abnormal in 87% of patients with ground glass opacities or local or patchy "shadowing".

Disease Progression

The more severe manifestations, such as ARDS, acute kidney injury (AKI) and myocardial injury, tend to occur as many as 8-14 days after the onset of symptoms and portend worse outcomes. Within a hospitalized population, rates of ICU admission range between 26-32% across most studies. Several studies have identified older age and baseline burden of comorbidity, such as diabetes, hypertension, prior coronary disease and prior lung disease, as predictors of more significant disease progression with higher rates of ARDS, AKI, cardiac injury, ICU admission and death.

- Increases in markers of inflammation, coagulation, and cardiac injury also correlate with disease severity and rise throughout the course of the disease.
- In hospitalized populations, the timing of death occurred at a median of 16-19 days after illness onset. The median time from symptom onset to discharge in survivors was around 3 weeks.
- Respiratory Failure

The most prominent complication of COVID-19 is respiratory failure. In hospitalized patients, respiratory symptoms are common and range in severity from cough (60-80%) or dyspnea (19-40%) to ARDS (17-42%).

- Renal injury

Estimates vary as to the incidence of acute kidney injury in COVID-19, ranging between 0.5-15.

- Liver injury

Transaminitis is common with an incidence of 21-37%, and as high as 48-62% of patients who are critically ill or who do not survive.

- Cardiac injury

Numerous studies have reported the acute cardiac injury as an important manifestation of COVID-19. The acute cardiac injury was variably defined as either cardiac troponin elevation >99th percentile alone or a composite of troponin elevation, ECG or echocardiographic abnormalities.

- Arrhythmia

Arrhythmias have been noted in several published reports. In a case series of 138 hospitalized patients with COVID-19, 16.7% (n=23) developed an unspecified arrhythmia during their hospitalization, higher rates were noted among patients admitted to the ICU (44.4%, n=16).

A case series of 187 hospitalized patients provided insight into specific arrhythmias, reporting sustained ventricular tachycardia or ventricular fibrillation amongst 5.9% (n=11) of the patients.

These findings are consistent with arrhythmias documented in influenza, which has been known to cause both AV node dysfunction and ventricular arrhythmias.

- Heart failure, cardiogenic shock and myocarditis

Heart failure and myocardial dysfunction have been described in COVID-19. In a case series of 191 patients, heart failure was noted as a complication of COVID-19 in 23% (n=44) of all patients and among 52% (n=28) of non-survivors.

- Thrombosis

One of the prominent findings replicated across most early studies of COVID-19 includes disarray of the coagulation and fibrinolytic system. Hospitalized patients with moderate and severe COVID-19 and those with poorer outcomes are noted to have prolonged prothrombin time (PT), elevated D-dimer, and activated partial thromboplastin time (APTT).

Mortality

Fatality rate estimates have been challenging with SARS-CoV2, as populations have not been widely screened for infection – leading to an underestimate of the denominator and probable overestimate of the CFR. Crude, unadjusted estimates for the global CFR are ~5% at present time with notable variation by country: Italy 11.9% (13,155 deaths), Spain 9.0% (9,387 deaths), South Korea 1.7% (169 deaths), China 4.1% (3312 deaths), Iran 6.4% (3036 deaths), Germany 1.2% (931 deaths) and the United States 2.3% (5,137 deaths).

Importantly, as healthcare capacity is exceeded, a large number of deaths may occur because of the limited availability of critical care resources, such as mechanical ventilation. When adjusted for underlying demography and under-ascertainment of cases, the CFR rate was estimated to be 1.4% in China.

- According to the Chinese CDC report of over 70,000 cases, the age-related CFR was as follows: <1% in age <50 years, 1.3% in age 50-59 years, 3.6% in age 60-69 years, 8% in age 70-79 years, and 14.8% in age 80 years and greater.
- This steep increase in age-related mortality was also observed in Italy, the US and South Korea. In fact, age, along with markers of disease severity (d-dimer and sequential organ failure assessment [SOFA] score) were the only independent predictors of mortality in one study.
- The presence of ACE2 receptors on the myocardium and vascular endothelial cells provides a theoretical mechanism for direct viral infection of the heart with resultant myocarditis.

Treatments in SARS CoV2

Nucleotide Analogs - Inhibitors of Viral Genome Replication:

- The antiviral mechanism of nucleotide analogs is to interfere with RdRp function and viral genome replication and amplification.
- There are no CoV-specific drugs available at this time and so ongoing efforts to employ this drug class against SARS-CoV2 are reliant on pre-existing agents designed for other viral illnesses.
- The most widely-applied agent in this class against SARS-CoV2 has been remdesivir. Remdesivir functions as a chain terminator of RNA replication initially designed for use against Ebola.
- Addition of remdesivir to the growing RNA strand by RdRp blocks the incorporation of additional nucleotides, thereby halting genome replication. The agent has been shown to have *in vitro* activity against SARS-CoV2, leading to off-label and investigational use around the world.

Another nucleotide analog for the disruption of RdRp-dependent viral replication is favipiravir, which has investigational approval in several countries. Additional agents that are under study include emtricitabine/tenofovir and ribavirin.

Protease Inhibitors - Inhibitors of Nonstructural Protein Generation:

- The antiviral mechanism of action of protease inhibitors is to block viral proteases which cleave the non-structural proteins from the large, monomeric, replicase as detailed above. As the maturation of non-structural proteins, such as RdRp, is necessary for viral reproduction, the pharmacologic impairment of the protease should be effective to stop viral replication.
- The trial of lopinavir-ritonavir, a combination protease inhibitor designed for HIV treatment, in 199 patients with at least moderate COVID-19 did not significantly alter clinical improvement or viral clearance. While the results of this trial were met with disappointment, this negative study should not forestall trials and drug development of protease inhibitors as a therapeutic class, given that this drug was not specifically designed for SARS-CoV2.
- The development of inhibitors specific to SARS-CoV2 main protease is underway. A class of agents identified using structure-based drug design, α-ketonamide inhibitors, has demonstrated *in vitro* efficacy and favorable pharmacokinetics. Other candidate protease inhibitors for SARS-CoV2 include danoprevir, a drug originally intended for HCV therapy.

Inhibitors of Membrane Fusion

In order for the viral genome to gain access to cellular machinery for replication, a membrane fusion event must occur between the viral and endosomal membranes, which are non-covalently bound by the interaction between the S protein and ACE2. The exact mechanism of membrane fusion is unknown but appears to be dependent on endosomal maturation and a membrane-bound host protease, TMPRSS2.

Chloroquine (CQ) / Hydroxychloroquine (HCQ)

- The antiviral properties of chloroquine (CQ) were previously observed in HIV and other viruses.
- CQ and HCQ are thought to inhibit endosomal maturation, a process by which endosomes are translocated from the perimembrane regions of the cell to central hubs (Figure 1). CQ prevented viral replication of SARS-CoV1 *in vitro*.
- The comparable efficacy of HCQ, a less toxic derivative, and suggested that the mechanism of impaired endosomal maturation indeed applied to SARS-CoV2 infection *in vitro*.
- Only poor-quality, non-randomized, unblinded data exists assessing the benefit of HCQ in COVID-19.
- While HCQ is being widely used with an FDA emergency authorization, more data are needed to prove efficacy against SARS-CoV2 in humans.

Camostat

- Camostat is a protease inhibitor approved for the treatment of chronic pancreatitis. Camostat appears to inhibit TMPRSS2 in proteomic and *in vitro* studies.
- A randomized, placebo-controlled trial is underway for this agent in COVID-19 (NCT04321096).

Neutralizing Antibodies and Decoy Proteins:

- Neutralizing antibodies are designed to bind virions, preventing viral exposure or binding to host cells.
- Plasma from patients who have recovered from SARS-CoV2 may contain anti-SARS-CoV2 IgG antibodies.
- In a small, single-arm trial of convalescent plasma in COVID-19 patients with ARDS, all had clinical improvement with 3/5 patients weaning from the ventilator. Additional trials are ongoing to better define the safety and efficacy of this strategy.
- Isolation of SARS-CoV2 specific neutralizing antibodies with clonal techniques is an appealing strategy to provide targeted therapy, potentially with a lower risk of adverse events.
- Strategies currently under investigation include antibodies cloned

from the convalescent serum of individuals recovered from SARS-CoV2 or SARS-CoV1 and synthetic antibodies.

- It is unclear whether differences in the S proteins of SARS-CoV1 and SARS-CoV2 may limit the effectiveness of antibodies cloned from patients convalescent to SARS-CoV1 (9). Synthetic antibodies represent an exciting, novel therapeutic avenue.
- One strategy being explored is to fuse ACE2 to Fc IgG, with the hypothesis that this synthetic antibody would serve as a decoy receptor, preventing cellular binding of the virion.

Anti-Inflammatory Therapy

- Advanced stages of COVID-19 have been likened to cytokine storm syndromes with non-specific widespread immune activation.
- Elevated levels of inflammatory biomarkers, such as IL-6 and hsCRP, identify patients at high risk of progressing to severe disease and death.
- Immunomodulatory and anti-inflammatory therapy has been used, despite limited data, in patients with evidence of hyper inflammation in an effort to curb pathologic immune activation.

Corticosteroids

- Corticosteroids have been used in several, severe viral respiratory infections including influenza, SARS-CoV and MERS-CoV with limited benefit and, in some instances, evidence of delayed viral clearance and increased rates of secondary infection and mortality.
- A retrospective analysis of 84 patients with ARDS secondary to SARS-CoV2 observed an association with improved survival in patients who received solumedrol.
- In the absence of robust evidence, major professional society guidelines do not recommend routine use of corticosteroids in the treatment of COVID-19 but rather restricting its use to patients with other indications for steroids, such as refractory shock or advanced ARDS.
- Clinical trials are ongoing to examine the safety and efficacy of corticosteroids in hospitalized non-critically ill COVID-19 patients (Clinical Trials).

IL-6 Inhibitors

- Elevation of IL-6 in patients with severe COVID-19 has prompted consideration of the use of IL-6 inhibitors (Tocilizumab, Siltuximab) extrapolating from the treatment of cytokine release syndrome.
- Tociluzimab, a recombinant humanized monoclonal antibody, and Siltuximab, a chimeric monoclonal antibody, both bind soluble and membrane-bound IL-6 receptors resulting in inhibition of IL-6-mediated signaling.
- In one preprint case series from China, 21 patients with severe or critical COVID-19 treated with tocilizumab experienced a salutary effect with a resolution of fever, improved oxygenation, improvement in lung opacities on chest CT, resolution of lymphopenia and a reduction in CRP levels within a few days of therapy in the absence of any significant reported adverse events.
- Several randomized clinical trials of tocilizumab in the treatment of severe COVID-19 infection are ongoing.

Azithromycin

- Azithromycin, a macrolide antibiotic, has long been touted for its anti-inflammatory effect and has been used as adjunctive therapy in the treatment of community-acquired pneumonia and chronic obstructive pulmonary disease exacerbations.
- Limited data suggest that adjunctive azithromycin in moderate-severe ARDS is associated with improved outcomes.
- A small non-randomized study showed that combination azithromycin and hydroxychloroquine was associated with more effective SARS-CoV2 clearance in COVID-19 patients compared with either monotherapy with hydroxychloroquine or standard of care; however, numerous limitations.
- Several randomized clinical trials assessing the combination of hydroxychloroquine/chloroquine with azithromycin across the severity spectrum of COVID-19 are ongoing or about to be launched.

Other Anti-Inflammatory Therapies

- JAK-2 inhibitors inhibit receptor mediated-endocytosis leading to the hypothesis that it might prevent the cellular entry of the SARS-CoV2. Additionally, this class of agents has anti-inflammatory effects by inhibiting cytokine release. An agent in the class, baricitinib, is being studied in an open-label non-randomized pilot study in patients with COVID-19 (NCT04320277).
- Currently, a 3-arm randomized control trial is being launched to compare anakinra monotherapy, emapalumab monotherapy and standard of care.
- Anakinra is a recombinant monoclonal antibody that blocks IL-1 receptors. It has been used to treat autoimmune conditions including juvenile idiopathic arthritis as well as recurrent pericarditis. Emapalumab is a fully human anti-gamma interferon monoclonal antibody that has been approved by the FDA for the treatment of primary hemophagocytic lymphohistiocytosis, a disease reminiscent of the hyperinflammatory state seen in advanced COVID-19. Finally, colchicine, a microtubule polymerization inhibitor and anti-inflammatory drug, is being tested in a large randomized clinical trial of ambulatory COVID-19 patients.
- *Angiotensin Converting Enzyme Inhibitors & Angiotensin Receptor Blockers*
- ACE-2 receptor mediated endocytosis of SARS-CoV2 is central to the viral life cycle. Conflicting data exist regarding the effect of renin-angiotensin-aldosterone-inhibitors (RAASi), including ACE-I and ARB, on ACE2 activity/levels in various human tissues and the resultant susceptibility to infection with SARS-CoV2 (18).
- The totality of the available data is insufficient to recommend cessation of ACE-I/ARB in individuals with an existing indication for life-prolonging therapy with these drugs and major societies have strongly recommended continuation of ACE-I/ARB therapy.
- An open-label randomized trial is on the way to examine the effect of prophylactic ACE-I/ARB withdrawal in COVID-19 naive individuals with essential hypertension as the sole indication for treatment on the risk of infection and subsequent complications.

- Based on the pre-clinical data described earlier in this review, two paired trials are currently underway examining losartan therapy in COVID-19 who are ambulatory and hospitalized.
- *Statins*
- The anti-inflammatory pleiotropic effects of statins have been cultivated in different pathologic states. Statins have been shown in murine models of acute lung injury and in humans to attenuate the inflammatory component of acute lung injury.
- A multi-center randomized trial of simvastatin in patients with versatile causes of ARDS showed no difference as compared to placebo in ventilator-free days, multi-organ failure and mortality.
- A subsequent study, sub phenotyping the trial population into hyper vs. hypoinflammatory ARDS, found a statistically significant improvement in survival with simvastatin in the hyperinflammatory group.
- A post-hoc analysis of the JUPITER trial observed a reduction in incident pneumonia with rosuvastatin. The benefit of statin therapy in the hyperinflammatory state in advanced COVID-19 is unknown.
- *Vaccines Against SARS-CoV2*
- As the discovery of a safe and efficacious vaccine again SARS-CoV2 is clearly the aspiration for preventative strategies, intense efforts are ongoing employing numerous approaches with accelerated testing.
- It is believed that all 4 structural proteins, E, M, N and S proteins, may serve as antigens for neutralizing antibody and CD4+CD8+ T cell responses.
- Based on the experience with SARS-CoV1 vaccine development, it seems that the most promising candidates target the S protein, which induces humoral and protective cellular immunity.
- Encouragingly, administration of full-length or the ACE2-receptor binding domain of the S protein of SARS-CoV1 induced highly potent neutralizing antibodies that conveyed protective immunity in animal models.
- Potential delivery strategies include inactivated or attenuated virus, subunit vaccines, viral vectors, DNA or RNA-based vaccines. Live attenuated viral vaccines are likely to induce

significant immune response but may carry the risk of disease, particularly in immunosuppressed individuals.

- Inactivated "whole" viral or subunit vaccines are relatively easy to develop, but do not induce immediate or complete immunity, typically requiring multiple doses to promote humoral, but often not cellular, immunity.
- Immunity may also wane over time, requiring booster dosing. Viral vector-based vaccines would employ other viruses, such as the vaccina virus (a poxvirus used for the smallpox vaccine) or adenovirus, to display SARS-CoV2 antigens.
- This strategy can promote robust cytotoxic T cell responses but may fail in the face of the pre-existing immunity to or toxicity of the viral vector.
- Nucleic acid-based strategies, which work through the delivery of DNA or RNA that are translated by host machinery to produce viral protein antigens, are relatively simple to design but may be limited by toxicity and/or stability concerns.
- Of note, at this time, there are no approved DNA or RNA vaccines for humans. Most approaches to SARS-CoV2 are in pre-clinical development, with several early trials of RNA (NCT04283461) and viral-vector vaccine strategies ongoing.

REFERENCES

1. Fehr AR, Perlman S. Coronaviruses: an overview of their replication and pathogenesis. Methods in molecular biology 2015;1282:1-23.
2. Zhu N, Zhang D, Wang W et al. A Novel Coronavirus from Patients with Pneumonia in China, 2019. The New England journal of medicine 2020;382:727-733.
3. Cui J, Li F, Shi ZL. Origin and evolution of pathogenic coronaviruses. Nature reviews Microbiology 2019;17:181-192.
4. Zhou P, Yang XL, Wang XG et al. A pneumonia outbreak associated with a new coronavirus of probable bat origin. Nature 2020;579:270-273.
5. Wu F, Zhao S, Yu B et al. A new coronavirus associated with human respiratory disease in China. Nature 2020;579:265-269.
6. Prentice E, McAuliffe J, Lu X, Subbarao K, Denison MR. Identification and characterization of severe acute respiratory syndrome coronavirus replicase proteins. Journal of virology 2004;78:9977-86.
7. Hoffmann M, Kleine-Weber H, Schroeder S et al. SARS-CoV-2 Cell Entry Depends on ACE2 and TMPRSS2 and Is Blocked by a Clinically Proven Protease Inhibitor. Cell 2020.
8. Du L, He Y, Zhou Y, Liu S, Zheng BJ, Jiang S. The spike protein of SARS-CoV--a target for vaccine and therapeutic development. Nature reviews Microbiology 2009;7:226-36.
9. Wrapp D, Wang N, Corbett KS et al. Cryo-EM structure of the 2019-nCoV spike in the prefusion conformation. Science 2020;367:1260-1263.
10. Yan R, Zhang Y, Li Y, Xia L, Guo Y, Zhou Q. Structural basis for the recognition of SARS-CoV-2 by full-length human ACE2. Science 2020;367:1444-1448.
11. Shang J, Ye G, Shi K et al. Structural basis of receptor recognition by SARS-CoV-2. Nature 2020.
12. Siu YL, Teoh KT, Lo J et al. The M, E, and N structural proteins of the severe acute respiratory syndrome coronavirus are required for efficient assembly, trafficking, and release of virus-like particles. Journal of virology 2008;82:11318-30.
13. Walls AC, Park YJ, Tortorici MA, Wall A, McGuire AT, Veesler D. Structure, Function, and Antigenicity of the SARS-CoV-2 Spike Glycoprotein. Cell 2020.

14. Bosch BJ, Smits SL, Haagmans BL. Membrane ectopeptidases targeted by human coronaviruses. Current opinion in virology 2014;6:55-60.
15. Cheng A, Zhang W, Xie Y et al. Expression, purification, and characterization of SARS coronavirus RNA polymerase. Virology 2005;335:165-76.
16. Li W, Moore MJ, Vasilieva N et al. Angiotensin-converting enzyme 2 is a functional receptor for the SARS coronavirus. Nature 2003;426:450-4.
17. Dijkman R, Jebbink MF, Deijs M et al. Replication-dependent downregulation of cellular angiotensin-converting enzyme 2 protein expression by human coronavirus NL63. The Journal of general virology 2012;93:1924-1929.
18. Vaduganathan M, Vardeny O, Michel T, McMurray JJV, Pfeffer MA, Solomon SD. Renin-Angiotensin-Aldosterone System Inhibitors in Patients with Covid-19. The New England journal of medicine 2020.
19. Clerkin KJ, Fried JA, Raikhelkar J et al. Coronavirus Disease 2019 (COVID-19) and Cardiovascular Disease. Circulation 2020.
20. Brosnihan KB, Neves LA, Chappell MC. Does the angiotensin-converting enzyme (ACE)/ACE2 balance contribute to the fate of angiotensin peptides in programmed hypertension? Hypertension 2005;46:1097-9.
21. Tikellis C, Thomas MC. Angiotensin-Converting Enzyme 2 (ACE2) Is a Key Modulator of the Renin Angiotensin System in Health and Disease. International journal of peptides 2012;2012:256294.
22. Hamming I, Cooper ME, Haagmans BL et al. The emerging role of ACE2 in physiology and disease. The Journal of pathology 2007;212:1-11.
23. Vickers C, Hales P, Kaushik V et al. Hydrolysis of biological peptides by human angiotensin-converting enzyme-related carboxypeptidase. The Journal of biological chemistry 2002;277:14838-43.
24. Hamming I, Timens W, Bulthuis ML, Lely AT, Navis G, van Goor H. Tissue distribution of ACE2 protein, the functional receptor for SARS coronavirus. A first step in understanding SARS pathogenesis. The Journal of pathology 2004;203:631-7.
25. Wang C, Horby PW, Hayden FG, Gao GF. A novel coronavirus outbreak of global health concern. Lancet 2020;395:470-473.
26. Imai Y, Kuba K, Rao S et al. Angiotensin-converting enzyme 2 protects from severe acute lung failure. Nature 2005;436:112-6.

27. Yang P, Gu H, Zhao Z et al. Angiotensin-converting enzyme 2 (ACE2) mediates influenza H7N9 virus-induced acute lung injury. Scientific reports 2014;4:7027.
28. Zou Z, Yan Y, Shu Y et al. Angiotensin-converting enzyme 2 protects from lethal avian influenza A H5N1 infections. Nature communications 2014;5:3594.
29. Kuba K, Imai Y, Rao S et al. A crucial role of angiotensin converting enzyme 2 (ACE2) in SARS coronavirus-induced lung injury. Nature medicine 2005;11:875-9.
30. Crackower MA, Sarao R, Oudit GY et al. Angiotensin-converting enzyme 2 is an essential regulator of heart function. Nature 2002;417:822-8.
31. Lambert DW, Yarski M, Warner FJ et al. Tumor necrosis factor-alpha convertase (ADAM17) mediates regulated ectodomain shedding of the severe-acute respiratory syndrome-coronavirus (SARS-CoV) receptor, angiotensin-converting enzyme-2 (ACE2). The Journal of biological chemistry 2005;280:30113-9.
32. Gu H, Xie Z, Li T et al. Angiotensin-converting enzyme 2 inhibits lung injury induced by respiratory syncytial virus. Scientific reports 2016;6:19840.
33. Khan A, Benthin C, Zeno B et al. A pilot clinical trial of recombinant human angiotensin-converting enzyme 2 in acute respiratory distress syndrome. Critical care 2017;21:234.
34. Monteil V KH, Prado P, Hagelkruys A, Wimmer RA, Stahl M, Leopoldi A, Garreta E, Hurtado del Pozo C, Prosper F, Romero JP, Wirnsberger G, Zhang H, Slutsky AS, Conder R, Montserrat N, Mirazimi A, Penninger J. Inhibition of SARS-CoV-2 infections in engineered human tissues using clinical-grade soluble human ACE2. Cell 2020.
35. Huang C, Wang Y, Li X et al. Clinical features of patients infected with 2019 novel coronavirus in Wuhan, China. Lancet 2020;395:497-506.
36. Lu R, Zhao X, Li J et al. Genomic characterisation and epidemiology of 2019 novel coronavirus: implications for virus origins and receptor binding. Lancet 2020;395:565-574.
37. Chan JF, Yuan S, Kok KH et al. A familial cluster of pneumonia associated with the 2019 novel coronavirus indicating person-to-person transmission: a study of a family cluster. Lancet 2020;395:514-523.
38. van Doremalen N, Bushmaker T, Morris DH et al. Aerosol and Surface Stability of SARS-CoV-2 as Compared with SARS-CoV-1. The New England journal of Medicine 2020.

39. Moriarty LF, Plucinski MM, Marston BJ et al. Public Health Responses to COVID-19 Outbreaks on Cruise Ships - Worldwide, February-March 2020. MMWR Morb Mortal Wkly Rep 2020;69:347-352.
40. Wang W, Xu Y, Gao R et al. Detection of SARS-CoV-2 in Different Types of Clinical Specimens. Jama 2020.
41. Alhazzani W, Moller MH, Arabi YM et al. Surviving Sepsis Campaign: Guidelines on the Management of Critically Ill Adults with Coronavirus Disease 2019 (COVID-19). Critical care medicine 2020.
42. Li R, Pei S, Chen B et al. Substantial undocumented infection facilitates the rapid dissemination of novel coronavirus (SARS-CoV2). Science 2020.
43. Callaway E, Cyranoski D, Mallapaty S, Stoye E, Tollefson J. The coronavirus pandemic in five powerful charts. Nature 2020;579:482-483.
44. Wei WE LZ, Chiew CJ, Yong SE, Toh MP, Lee VJ. Presymptomatic Transmission of SARS-CoV-2 Singapore, January 23–March 16, 2020. MMWR Morb Mortal Wkly Rep
45. Wu Z, McGoogan JM. Characteristics of and Important Lessons From the Coronavirus Disease 2019 (COVID-19) Outbreak in China: Summary of a Report of 72314 Cases From the Chinese Center for Disease Control and Prevention. Jama 2020.
46. Centers for Disease C. Severe Outcomes Among Patients with Coronavirus Disease 2019 (COVID-19) — United States. MMWR Morb Mortal Wkly Rep 2020;69:343-346.
47. Lipsitch M, Swerdlow DL, Finelli L. Defining the Epidemiology of Covid-19 - Studies Needed. The New England journal of medicine 2020;382:1194-1196.
48. Guan WJ, Ni ZY, Hu Y et al. Clinical Characteristics of Coronavirus Disease 2019 in China. The New England journal of medicine 2020.
49. Surgery AAoO-HaN. Lauer SA, Grantz KH, Bi Q et al. The Incubation Period of Coronavirus Disease 2019 (COVID-19) From Publicly Reported Confirmed Cases: Estimation and Application. Ann Intern Med 2020.
51. Wu C, Chen X, Cai Y et al. Risk Factors Associated With Acute Respiratory Distress Syndrome and Death in Patients With Coronavirus Disease 2019 Pneumonia in Wuhan, China. JAMA Intern Med 2020.
52. Wang D, Hu B, Hu C et al. Clinical Characteristics of 138 Hospitalized Patients With 2019 Novel Coronavirus-Infected Pneumonia in

Wuhan, China. Jama 2020.
53. Zhou F, Yu T, Du R et al. Clinical course and risk factors for mortality of adult inpatients with COVID-19 in Wuhan, China: a retrospective cohort study. Lancet 2020;395:1054-1062.
54. Guo T, Fan Y, Chen M et al. Cardiovascular Implications of Fatal Outcomes of Patients With Coronavirus Disease 2019 (COVID-19). JAMA Cardiol 2020.
55. Arentz M, Yim E, Klaff L et al. Characteristics and Outcomes of 21 Critically Ill Patients With COVID-19 in Washington State. Jama 2020.
56. Shi S, Qin M, Shen B et al. Association of Cardiac Injury With Mortality in Hospitalized Patients With COVID-19 in Wuhan, China. JAMA Cardiol 2020.
57. Chen N, Zhou M, Dong X et al. Epidemiological and clinical characteristics of 99 cases of 2019 novel coronavirus pneumonia in Wuhan, China: a descriptive study. Lancet 2020;395:507-513.
58. Ruan Q, Yang K, Wang W, Jiang L, Song J. Clinical predictors of mortality due to COVID-19 based on an analysis of data of 150 patients from Wuhan, China. Intensive Care Med 2020.
59. Cheng Y LR, Wang K, Zhang M, Wang Z, Dong L, Li J, Yao Y, Ge S, Xu G. . Kidney disease is associated with in-hospital death of patients with COVID-19. Kidney International 2020.
60. Inciardi RM, Lupi L, Zaccone G et al. Cardiac Involvement in a Patient With Coronavirus Disease 2019 (COVID-19). JAMA Cardiol 2020.
61. Fleet JC, Clinton SK, Salomon RN, Loppnow H, Libby P. Atherogenic diets enhance endotoxin-stimulated interleukin-1 and tumor necrosis factor gene expression in rabbit aortae. J Nutr 1992;122:294-305.
62. Estabragh ZR, Mamas MA. The cardiovascular manifestations of influenza: a systematic review. Int J Cardiol 2013;167:2397-403.
63. Hu H, Ma F, Wei X, Fang Y. Coronavirus fulminant myocarditis saved with glucocorticoid and human immunoglobulin. Eur Heart J 2020.
64. Deng Y, Liu W, Liu K et al. Clinical characteristics of fatal and recovered cases of coronavirus disease 2019 (COVID-19) in Wuhan, China: a retrospective study. Chin Med J (Engl) 2020.
65. Hwang DM, Chamberlain DW, Poutanen SM, Low DE, Asa SL, Butany J. Pulmonary pathology of severe acute respiratory syndrome in Toronto. Mod Pathol 2005;18:1-10.

66. Luo WY, H.; Gou, J.; Li, X.; Sun, Y.; Li, J.; Liu, L. Clinical Pathology of Critical Patient with Novel Coronavirus Pneumonia (COVID-19). Preprints 2020.
67. Hematology ASo. COVID-19 and Coagulopathy: Frequently Asked Questions. 2020.
68. Ruan S. Likelihood of survival of coronavirus disease 2019. Lancet Infect Dis 2020. Medicine JHU. Coronavirus Resource Center. 2020.
70. Verity R, Okell LC, Dorigatti I et al. Estimates of the severity of coronavirus disease 2019: a model-based analysis. Lancet Infect Dis 2020.
71. Onder G, Rezza G, Brusaferro S. Case-Fatality Rate and Characteristics of Patients Dying in to COVID-19 in Italy. Jama 2020.
72. Dudley JP, Lee NT. Disparities in Age-Specific Morbidity and Mortality from SARS-CoV-2 in China and the Republic of Korea. Clin Infect Dis 2020.
73. Kumar A, Thota V, Dee L, Olson J, Uretz E, Parrillo JE. Tumor necrosis factor alpha and interleukin 1beta are responsible for *in vitro* myocardial cell depression induced by human septic shock serum. J Exp Med 1996;183:949-58.
74. Oudit GY, Kassiri Z, Jiang C et al. SARS-coronavirus modulation of myocardial ACE2 expression and inflammation in patients with SARS. Eur J Clin Invest 2009;39:618-25.
75. Ding Y, He L, Zhang Q et al. Organ distribution of severe acute respiratory syndrome (SARS) associated coronavirus (SARS-CoV) in SARS patients: implications for pathogenesis and virus transmission pathways. The Journal of pathology 2004;203:622-30.
76. To KF, Tong JH, Chan PK et al. Tissue and cellular tropism of the coronavirus associated with severe acute respiratory syndrome: an in-situ hybridization study of fatal cases. The Journal of pathology 2004;202:157-63.
77. Ding Y, Wang H, Shen H et al. The clinical pathology of severe acute respiratory syndrome (SARS): a report from China. The Journal of pathology 2003;200:282-9.
78. Guillevin L. Virus-induced systemic vasculitides: new therapeutic approaches. Clin Dev Immunol 2004;11:227-31.
79. Pagnoux C, Cohen P, Guillevin L. Vasculitides secondary to infections. Clin Exp Rheumatol 2006;24:S71-81.
80. Ding YQ, Wang HJ, Shen H et al. [Study on etiology and pathology of severe acute respiratory syndrome]. Zhonghua Bing Li Xue Za Zhi

2003;32:195-200.
81. Tang N, Li D, Wang X, Sun Z. Abnormal coagulation parameters are associated with poor prognosis in patients with novel coronavirus pneumonia. J Thromb Haemost 2020;18:844-847.
82. Levi M, van der Poll T, ten Cate H, van Deventer SJ. The cytokine-mediated imbalance between coagulant and anticoagulant mechanisms in sepsis and endotoxaemia. Eur J Clin Invest 1997;27:3-9.
83. Simmons J, Pittet JF. The coagulopathy of acute sepsis. Curr Opin Anaesthesiol 2015;28:227-36.
84. Nakamura S, Imamura T, Okamoto K. Tissue factor in neutrophils: yes. J Thromb Haemost 2004;2:214-7.
85. van der Poll T, Buller HR, ten Cate H et al. Activation of coagulation after administration of tumor necrosis factor to normal subjects. The New England journal of medicine 1990;322:1622-7.
86. de Jonge E, Friederich PW, Vlasuk GP et al. Activation of coagulation by administration of recombinant factor VIIa elicits interleukin 6 (IL-6) and IL-8 release in healthy human subjects. Clin Diagn Lab Immunol 2003;10:495-7.
87. Franco RF, de Jonge E, Dekkers PE et al. The in vivo kinetics of tissue factor messenger RNA expression during human endotoxemia: relationship with activation of coagulation. Blood 2000;96:554-9.
88. Levi M, van der Poll T, Buller HR. Bidirectional relation between inflammation and coagulation. Circulation 2004;109:2698-704.
89. Della Valle P, Pavani G, D'Angelo A. The protein C pathway and sepsis. Thromb Res 2012;129:296-300.
90. Green J, Doughty L, Kaplan SS, Sasser H, Carcillo JA. The tissue factor and plasminogen activator inhibitor type-1 response in pediatric sepsis-induced multiple organ failure. Thromb Haemost 2002;87:218-23.
91. Cox D, Kerrigan SW, Watson SP. Platelets and the innate immune system: mechanisms of bacterial-induced platelet activation. J Thromb Haemost 2011;9:1097-107.
92. Gawaz M, Dickfeld T, Bogner C, Fateh-Moghadam S, Neumann FJ. Platelet function in septic multiple organ dysfunction syndrome. Intensive Care Med 1997;23:379-85.
93. Akca S, Haji-Michael P, de Mendonca A, Suter P, Levi M, Vincent JL. Time course of platelet counts in critically ill patients. Critical care medicine 2002;30:753-6.

94. Lee KH, Hui KP, Tan WC. Thrombocytopenia in sepsis: a predictor of mortality in the intensive care unit. Singapore Med J 1993;34:245-6.
95. Medina de Chazal H, Del Buono MG, Keyser-Marcus L et al. Stress Cardiomyopathy Diagnosis and Treatment: JACC State-of-the-Art Review. J Am Coll Cardiol 2018;72:1955-1971.
96. Corrales-Medina VF, Alvarez KN, Weissfeld LA et al. Association between hospitalization for pneumonia and subsequent risk of cardiovascular disease. Jama 2015;313:264-74.
97. Udell JA, Zawi R, Bhatt DL et al. Association between influenza vaccination and cardiovascular outcomes in high-risk patients: a meta-analysis. Jama 2013;310:1711-20.
98. Kwong JC, Schwartz KL, Campitelli MA et al. Acute Myocardial Infarction after Laboratory-Confirmed Influenza Infection. The New England journal of medicine 2018;378:345-353.
99. Welt FGP, Shah PB, Aronow HD et al. Catheterization Laboratory Considerations During the Coronavirus (COVID-19) Pandemic: From ACC's Interventional Council and SCAI. J Am Coll Cardiol 2020.
100. Libby P, Loscalzo J, Ridker PM et al. Inflammation, Immunity, and Infection in Atherothrombosis: JACC Review Topic of the Week. J Am Coll Cardiol 2018;72:2071-2081.
101. Ridker PM, Everett BM, Thuren T et al. Antiinflammatory Therapy with Canakinumab for Atherosclerotic Disease. The New England journal of medicine 2017;377:1119-1131.
102. Libby P. Inflammation in atherosclerosis. Arterioscler Thromb Vasc Biol 2012;32:2045-51.
103. Violi F, Cangemi R, Calvieri C. Pneumonia, thrombosis and vascular disease. J Thromb Haemost 2014;12:1391-400.
104. Mogensen TH. Pathogen recognition and inflammatory signaling in innate immune defenses. Clin Microbiol Rev 2009;22:240-73, Table of Contents.
105. van de Veerdonk FL, Netea MG, Dinarello CA, Joosten LA. Inflammasome activation and IL-1beta and IL-18 processing during infection. Trends Immunol 2011;32:110-6.
106. Vallance P, Collier J, Bhagat K. Infection, inflammation, and infarction: does acute endothelial dysfunction provide a link? Lancet 1997;349:1391-2.

107. Sarkisian L, Saaby L, Poulsen TS et al. Prognostic Impact of Myocardial Injury Related to Various Cardiac and Noncardiac Conditions. Am J Med 2016;129:506-514 e1.
108. Lim W, Qushmaq I, Devereaux PJ et al. Elevated cardiac troponin measurements in critically ill patients. Arch Intern Med 2006;166:2446-54.
109. Sarkisian L, Saaby L, Poulsen TS et al. Clinical Characteristics and Outcomes of Patients with Myocardial Infarction, Myocardial Injury, and Nonelevated Troponins. Am J Med 2016;129:446 e5-446 e21.
110. Thygesen K, Alpert JS, Jaffe AS et al. Fourth Universal Definition of Myocardial Infarction (2018). J Am Coll Cardiol 2018;72:2231-2264.
111. Chapman AR, Shah ASV, Lee KK et al. Long-Term Outcomes in Patients With Type 2 Myocardial Infarction and Myocardial Injury. Circulation 2018;137:1236-1245.
112. Qin C, Zhou L, Hu Z et al. Dysregulation of immune response in patients with COVID-19 in Wuhan, China. Clin Infect Dis 2020.
113. Grupp SA, Kalos M, Barrett D et al. Chimeric antigen receptor-modified T cells for acute lymphoid leukemia. The New England journal of medicine 2013;368:1509-1518.
114. Mehta P, McAuley DF, Brown M et al. COVID-19: consider cytokine storm syndromes and immunosuppression. Lancet 2020;395:1033-1034.
115. Siddiqi H, Mehra, MR. COVID-19 Illness in Native and Immunosuppressed States: A Clinical-Therapeutic Staging Proposal. The Journal of Heart and Lung Transplantation 2020.
116. Frey N, Porter D. Cytokine Release Syndrome with Chimeric Antigen Receptor T Cell Therapy. Biol Blood Marrow Transplant 2019;25:e123-e127.
117. Natanson C, Eichenholz PW, Danner RL et al. Endotoxin and tumor necrosis factor challenges in dogs simulate the cardiovascular profile of human septic shock. J Exp Med 1989;169:823-32.
118. Pathan N, Hemingway CA, Alizadeh AA et al. Role of interleukin 6 in myocardial dysfunction of meningococcal septic shock. Lancet 2004;363:203-9.
119. Goldhaber JI, Kim KH, Natterson PD, Lawrence T, Yang P, Weiss JN. Effects of TNF-alpha on [Ca2+]i and contractility in isolated adult rabbit ventricular myocytes. Am J Physiol 1996;271:H1449-55.
120. Krown KA, Yasui K, Brooker MJ et al. TNF alpha receptor expres-

sion in rat cardiac myocytes: TNF alpha inhibition of L-type Ca2+ current and Ca2+ transients. FEBS Lett 1995;376:24-30.
121. Hobai IA, Edgecomb J, LaBarge K, Colucci WS. Dysregulation of intracellular calcium transporters in animal models of sepsis-induced cardiomyopathy. Shock 2015;43:3-15.
122. Balligand JL, Ungureanu D, Kelly RA et al. Abnormal contractile function due to induction of nitric oxide synthesis in rat cardiac myocytes follows exposure to activated macrophage-conditioned medi um. J Clin Invest 1993;91:2314-9.
123. Stanzani G, Duchen MR, Singer M. The role of mitochondria in sepsis-induced cardiomyopathy. Biochim Biophys Acta Mol Basis Dis 2019;1865:759-773.
124. Mulangu S, Dodd LE, Davey RT, Jr. et al. A Randomized, Controlled Trial of Ebola Virus Disease Therapeutics. The New England journal of medicine 2019;381:2293-2303.
125. Holshue ML, DeBolt C, Lindquist S et al. First Case of 2019 Novel Coronavirus in the United States. The New England journal of medicine 2020;382:929-936.
126. Gordon CJ, Tchesnokov EP, Feng JY, Porter DP, Gotte M. The antiviral compound remdesivir potently inhibits RNA-dependent RNA polymerase from Middle East respiratory syndrome coronavirus. The Journal of biological chemistry 2020.
127. Sheahan TP, Sims AC, Leist SR et al. Comparative therapeutic efficacy of remdesivir and combination lopinavir, ritonavir, and interferon beta against MERS-CoV. Nature communications 2020;11:222.
128. Harrison C. Coronavirus puts drug repurposing on the fast track. Nature biotechnology 2020.
129. Dong L, Hu S, Gao J. Discovering drugs to treat coronavirus disease 2019 (COVID-19). Drug discoveries & therapeutics 2020;14:58-60.
130. Chan KW, Wong VT, Tang SCW. COVID-19: An Update on the Epidemiological, Clinical, Preventive and Therapeutic Evidence and Guidelines of Integrative Chinese-Western Medicine for the Management of 2019 Novel Coronavirus Disease. The American journal of Chinese medicine 2020:1-26.
131. Cao B, Wang Y, Wen D et al. A Trial of Lopinavir-Ritonavir in Adults Hospitalized with Severe Covid-19. The New England journal of medicine 2020.
132. Zhang L, Lin D, Sun X et al. Crystal structure of SARS-CoV-2 main protease provides a basis for design of improved alpha-ketoamide inhibitors. Science 2020.

133. Chen H ZZ, Wang L, Huang Z, Gong F, Li X, Chen Y, Wu JJ. First Clinical Study Using HCV Protease Inhibitor Danoprevir to Treat Naive and Experienced COVID-19 Patients. MedRxiv 2020.
134. Savarino A, Gennero L, Sperber K, Boelaert JR. The anti-HIV-1 activity of chloroquine. Journal of clinical virology : the official publication of the Pan American Society for Clinical Virology 2001;20:131-5.
135. Savarino A, Gennero L, Chen HC et al. Anti-HIV effects of chloroquine: mechanisms of inhibition and spectrum of activity. Aids 2001;15:2221-9.
136. Mauthe M, Orhon I, Rocchi C et al. Chloroquine inhibits autophagic flux by decreasing autophagosome-lysosome fusion. Autophagy 2018;14:1435-1455.
137. Scott CC, Vacca F, Gruenberg J. Endosome maturation, transport and functions. Seminars in cell & developmental biology 2014;31:2-10.
138. Vincent MJ, Bergeron E, Benjannet S et al. Chloroquine is a potent inhibitor of SARS coronavirus infection and spread. Virology journal 2005;2:69.
139. Liu J, Cao R, Xu M et al. Hydroxychloroquine, a less toxic derivative of chloroquine, is effective in inhibiting SARS-CoV-2 infection *in vitro*. Cell discovery 2020;6:16.
140. Gautret P, Lagier JC, Parola P et al. Hydroxychloroquine and azithromycin as a treatment of COVID-19: results of an open-label non-randomized clinical trial. International journal of antimicrobial agents 2020:105949.
141. Gordon DE JG, Bouhaddou M, Xu J, et al. A SARS-CoV-2-Human Protein-Protein Interaction Map Reveals Drug Targets and Potential Drug-Repurposing. BioRxiv 2020.
142. Shen C, Wang Z, Zhao F et al. Treatment of 5 Critically Ill Patients With COVID-19 With Convalescent Plasma. Jama 2020.
143. Lei C FW, Zian K, Li T, Zhang S, Ding M, Hu S. Potent neutralization of 2019 novel coronavirus by recombinant ACE2-Ig. BioRxiv 2020.
144. Russell CD, Millar JE, Baillie JK. Clinical evidence does not support corticosteroid treatment for 2019-nCoV lung injury. Lancet 2020;395:473-475.
145. Le RQ, Li L, Yuan W et al. FDA Approval Summary: Tocilizumab for Treatment of Chimeric Antigen Receptor T Cell-Induced Severe or Life-Threatening Cytokine Release Syndrome. The oncologist 2018;23:943-947.

146. Xu X HM, Li T, Sun W, Wang D, Fu B, Zhou Y, Zheng X, Yang Y, Li X, Zhang X, Pan A, Wei H. Effective Treatment of Severe COVID-19 Patients with Tocilizumab. ChinaXiv 2020.
147. Zimmermann P, Ziesenitz VC, Curtis N, Ritz N. The Immunomodulatory Effects of Macrolides-A Systematic Review of the Underlying Mechanisms. Frontiers in immunology 2018;9:302.
148. Kawamura K, Ichikado K, Takaki M, Eguchi Y, Anan K, Suga M. Adjunctive therapy with azithromycin for moderate and severe acute respiratory distress syndrome: a retrospective, propensity score-matching analysis of prospectively collected data at a single center. International journal of antimicrobial agents 2018;51:918-924.
149. Richardson P, Griffin I, Tucker C et al. Baricitinib as potential treatment for 2019-nCoV acute respiratory disease. Lancet 2020;395:e30-e31.
150. Jacobson JR, Barnard JW, Grigoryev DN, Ma SF, Tuder RM, Garcia JG. Simvastatin attenuates vascular leak and inflammation in murine inflammatory lung injury. American journal of physiology Lung cellular and molecular physiology 2005;288:L1026-32.
151. Shyamsundar M, McKeown ST, O'Kane CM et al. Simvastatin decreases lipopolysaccharide-induced pulmonary inflammation in healthy volunteers. American journal of respiratory and critical care medicine 2009;179:1107-14.
152. McAuley DF, Laffey JG, O'Kane CM et al. Simvastatin in the acute respiratory distress syndrome. The New England journal of medicine 2014;371:1695-703.
153. Calfee CS, Delucchi KL, Sinha P et al. Acute respiratory distress syndrome subphenotypes and differential response to simvastatin: secondary analysis of a randomised controlled trial. The Lancet Respiratory medicine 2018;6:691-698.
154. Novack V, MacFadyen J, Malhotra A, Almog Y, Glynn RJ, Ridker PM. The effect of rosuvastatin on incident pneumonia: results from the JUPITER trial. CMAJ 2012;184:E367-72.
155. Jiang S, He Y, Liu S. SARS vaccine development. Emerging infectious diseases 2005;11:1016-20.
156. Bukreyev A, Lamirande EW, Buchholz UJ et al. Mucosal immunisation of African green monkeys (Cercopithecus aethiops) with an attenuated parainfluenza virus expressing the SARS coronavirus spike protein for the prevention of SARS. Lancet 2004;363:2122-7.
157. Song Z, Xu Y, Bao L et al. From SARS to MERS, Thrusting Coronaviruses into the Spotlight. Viruses 2019;11.

158. Shang W, Yang Y, Rao Y, Rao X. The outbreak of SARS-CoV-2 pneumonia calls for viral vaccines. NPJ Vaccines 2020;5:18.
159. Ura T, Okuda K, Shimada M. Developments in Viral Vector-Based Vaccines. Vaccines (Basel) 2014;2:624-41.
160. Emanuel EJ, Persad G, Upshur R et al. Fair Allocation of Scarce Medical Resources in the Time of Covid-19. The New England journal of medicine 2020.
161. SIAARTI. Clinical Ethics Recommendations For the Allocation of Intensive Care Treatments, in exceptional, Resource-Limited Circumstances. 2020.
162. Crisis Standards of Care: A Systems Framework for Catastrophic Disaster Response. Washington (DC), 2012.
163. Biddison LD, Berkowitz KA, Courtney B et al. Ethical considerations: care of the critically ill and injured during pandemics and disasters: CHEST consensus statement. Chest 2014;146:e145S-55S.
164. Truog RD, Mitchell C, Daley GQ. The Toughest Triage - Allocating Ventilators in a Pandemic. The New England journal of medicine 2020.
165. Shahpori R, Stelfox HT, Doig CJ, Boiteau PJ, Zygun DA. Sequential Organ Failure Assessment in H1N1 pandemic planning. Critical care medicine 2011;39:827-32.
166. Khan Z, Hulme J, Sherwood N. An assessment of the validity of SOFA score based triage in H1N1 critically ill patients during an influenza pandemic. Anaesthesia 2009;64:1283-8.

PART-II

GENOME EDITING THROUGH CRISPR

CHAPTER 5

CRISPR/Cas TECHNOLOGY USE IN THERAPEUTICS APPLICATIONS

INTRODUCTION

Clustered regularly interspaced short palindromic repeats (CRISPR)-Cas system is a powerful tool for making changes to the genome of numerous life forms. Most CRISPR-Cas types have screens or performed with *in vitro* or in vivo. First found in quite a while as a component of a versatile resistant framework, CRISPR/Cas system changed variants have been discovered to design genomes and to pass or to smother the statement of qualities. In that capacity, the CRISPR/Cas system has involved in cell-based treatments. CRISPR has demonstrated a remarkable clinical potential to find novel focuses on treatment. Besides, the CRISPR-Cas system can be utilized to quickly design safe immunotherapeutic applications. Maybe more critically, the capacity of CRISPR-Cas to precisely alter qualities, not just in cell culture models and model life forms yet additionally in people, permits its utilization in restorative investigations. Therefore, in this review, we highlighted the CRISPR-Cas system has clinical potential in cell therapies and also discover new therapeutic agents.

CRISPR/Cas9 has a powerful technique for making changes to the genome of numerous living beings or different organisms. First found in microbes like bacteria as a component of a versatile resistant framework, CRISPR/Cas9 and changed variants have discovered a broad use to build genomes and to enact or to repress the expression qualities of the gene. So, CRISPR/Cas9 has enhanced the efficiency to know about the disease like cancer. Consequently, through this technology to find out the effective mechanism of tumorigenesis, so we can easily recognize or focuses on drug development, and conceivably also developed cell-based treatments. The field of genome altering has progressed at incredible speed, yielding effectively open, flexible, and reasonable method to consider in science. Specifically, the CRISPR framework has formed into a generally applied innovation quality capacity *in vitro* and *in vivo* system (Heidenrich and Zhang, 2016; Lucas et al., 2017; Biagiani et al., 2018; Yin et al., 2019; Liu et al., 2019).

The CRISPR genome altering tool, as we are aware it today, comprise of the CRISPR associated protein Cas9 isolated from *Streptococcus pyogenes* (SpCas9), which is an RNA-guided nuclease that is coordinated to a particular genomic area by a single guide RNA (sgRNA) (Jinek et al., 2012; Cong et al., 2013; Mali et al., 2013). The main essential for an appropriate official of the Cas-sgRNA complex is the nearness of a protospacer nearby PAM sequences that is perceived by the PAM-cooperating space of the Cas9 protein (Jinek et al., 2014; Stonberg et al., 2014).

After convincing, Cas9 makes a nick on the correlative and non-complementary DNA strand, in this way making a double-strand break (DSB). As of late, various reports have uncovered that Cas9 will in general leave a single nucleotide overhang four nucleotides distal of the PAM sequence, contradicting the general thought that a blunt cut is made (Zuo and Liu, 2016; Lemos et al., 2018). The DSB made by Cas9 will in this manner be fixed by endogenous mechanisms, for example, nonhomologous end-joining (NHEJ) or homology-coordinated repair (HDR) (Figure 1). Significantly, researchers have found that the HDR pathway can be utilized to add any genomic sequence of enthusiasm into the genome of the host by providing an exogenous DNA sequence with homology arms (Kim et al., 2014; Lin et al., 2014; Maryama et al., 2015).

In the recent decade, the CRISPR-Cas9 technology has converted genome engineering via disposing of the want for any understanding in engineering custom centered DNA-binding proteins due to the fact the goal specificity of CRISPR-Cas9 is based on the base pairing of nucleic acids as opposed to protein-DNA identification. CRISPR-Cas tool is a prokaryotic adaptive immunity mechanism used to cleave invading nucleic acids (Barrangou et al., 2007). A collection of CRISPR-Cas systems exist throughout numerous species of microorganisms like bacteria and archaea, which vary in their additives and mechanisms of action. For example, class 1 CRISPR-Cas systems comprise multiprotein effector complexes, while class 2 have a single effector protein usual there are six CRISPR-Cas kinds and as a minimum 29 subtypes (Makarova et al., 2015; Shmakov et al., 2017; Koonen et al., 2017), and this listing of sorts and subtypes is undergoing speedy enlargement.

All CRISPR-Cas structures depend upon CRISPR RNA (crRNA) or, in experimental CRISPR-Cas9 systems, on the guide RNA (gRNA) for guidance and targeting specificity. Following hybridization of the spacer part of the crRNA to a target sequence this is located subsequent to a protospacer adjoining motif (PAM) (or a protospacer flanking sequence (PFS) in type VI structures), the Cas cleaves the target nucleic acid. Thus, website-specific cleavage at any locus containing a PAM or a PFS may be completed through retargeting CRISPR-Cas structures with designed crRNAs containing appropriate spacer sequences. The discovery and development of type II CRISPR-Cas9 structures and the benefit of their use have brought about speedy adoption and development of a great range of applications, from fundamental technological know-how to translational science and medicine (Hsu et al., 2014). In turn, the early successes have stimulated efforts to discover new CRISPR-Cas systems and increase novel genome engineering programs.

We suggest the latest advances in CRISPR-Cas tools for gene editing and epigenetic modulation, before describing a diverse range of new CRISPR-Cas features and in healing capability transferring toward scientific studies (Figure 1). Here, we provide a detailed review of the examinations that have applied genome editing technique through CRISPR tool for better knowledge about the mechanism *in vitro* or *in vivo* system and also improvements that further enhance the utilization of the CRISPR tool to develop as a new therapeutic approach.

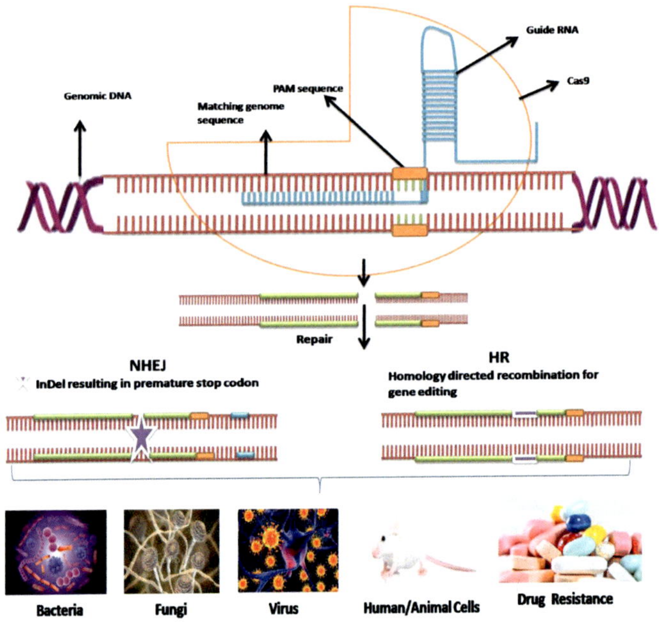

Figure 1. Clustered Regularly Interspaced Short Palindromic Repeats (CRISPR) may improve our understanding of other processes, such as Bacteria, Fungi, Virus, Human and Animal cells and Drug resistance.

GENOME EDITING

CRISPR-Cas structures are modular DNA-binding or RNA-binding proteins that can be engineered to bind specific sequences via designing crRNAs or gRNAs containing spacers that complement the target sequence. In addition to binding particular nucleic acid sequences, these proteins additionally characteristic as nucleases and thus may be used for programmable genome modifying. With the unique attributes of diverse CRISPR-Cas systems, along with PAM specificity, protein length and nuclease activity, quite a number CRISPR-Cas-primarily based DNA-targeting gear were evolved for genome editing packages. Additionally, the development of strategies for the detection of on-target and off-target interactions has advanced the concentration on the specificity of the CRISPR-Cas tool.

Proteins Involved in CRISPR-Cas Systems

Two proteins, Cas1 and Cas2, that are available in the incredibly larger part of the known CRISPR-Cas systems are adequate for the addition of spacers into the CRISPR cassettes (Yosef et al., 2012). These two proteins structure a confounding that is required for this adjustment procedure; the endonuclease action of Cas1 is required for spacer reconciliation through Cas2 seems to play out a non-enzymatic activity (Nunez et al., 2014; Richter et al., 2012). The Cas1-Cas2 complex speaks to the exceptionally preserved "data preparing" module of CRISPR-Cas that seems, by all accounts, to be semi self-sufficient from the remaining systems.

The subsequent stage, the handling of pre-crRNA into the guide crRNAs, is performed either by a devoted RNA endonuclease complex or by means of an elective system that includes bacterial RNase III and an extra RNA species (Deltcheva et al., 2011). The developed crRNA is bound by one (type II) or a few (types I and III) Cas proteins that structure the effector complex, which focuses on the related DNA or RNA (Hale et al., 2012; Jinek et al., 2012; Sashital et al., 2012; Van Duijn et al., 2012; Zhang et al., 2012; Wiedenheft et al., 2011). The effector complex of type I frameworks is known as Cascade (CRISPR-related complex for antiviral defense) (Brouns et al., 2008).

CRISPR-Cas Genes Contribute to Genome Editing

The three kinds are promptly discernable by uprightness of the nearness of three exceptional mark genes: Cas3 in type I systems, Cas9 in type II, and Cas10 in type III (Makarova et al., 2011). With a few special cases, every one of the three CRISPR-Cas types contains full parts that are required for the key steps of the defense system. As of late, on account of inside and out, grouping examination and structure of the effectors make from various variations of CRISPR-Cas systems, basic standards of association and capacity of the complexes have been revealed taking into account further overview of the CRISPR-Cas classification (Zhang et al., 2013; Staals et al., 2013; Spilman et al., 2013; Makarova et al., 2011).

Classification of Cas System

Cas1

Cas1 is a type I system in CRISPR-Cas frameworks and metal dependent deoxyribonuclease (Wiedenheft et al., 2009; Han et al., 2009); deletion of Cas1 in *E.coli* results in increased sensitivity to DNA damage and impaired the segregation of chromosomal. The available structures and the features of structures such as 3LFX, 2YZS, 3GOD and PDB which is a unique fold with two domains site (N-terminal β-stranded domain and catalytic C-terminal α-helical domain (Babu et al., 2011).

Cas2

Cas2 is type I systems encoded by a single operon containing the Cas2 gene. It is RNase specific to U-rich regions and double-stranded DNase. The available structures are 3EXC, RRM (Ferredoxin) fold ,PDB:21VY and 218E (Beloglazova et al., 2008; Nam et al., 2012).

Cas3

Cas3 is a type I system of the class 1 classification that are the most well-known kind of CRISPR-Cas frameworks in nature, involving a multimeric DNA-focusing on complex named 'Course' and the endonuclease Cas3. Before enlisting Cas3 to an objective DNA grouping, Cascade should initially bind to DNA through PAM and spacer recognition (Gao et al., 2017; Jore et al., 2017; Westra et al., 2012; Hochstrasser et al., 2014; Brendel et al., 2014; Richter et al., 2014; Wiedenheft et al., 2016). The course offers more noteworthy objective site adaptability attributable to its unbridled acknowledgment of PAM sequences 43. Enrollment of Cas3 produces a solitary stranded scratch, trailed by target DNA corruption through 3′ to 5′ exonuclease activity (Jore et al., 2017; Hochstrasser et al., 2014; Sinkunas et al., 2011; Jackson et al., 2014). Both the nickase movement and the helicase action of Cas3 are fundamental for the modification of remote DNA in prokaryotes (Westra et al., 2012). The exceptional cutting instrument of Cas3 is being saddled as an antimicrobial device by coordinating local or exogenous sort I systems to bacterial genomes for degradation and consequent cell death (Gomaa et al., 2014). Investigation to repurpose the nickase, helicase and exonuclease exercises of Cas3 may prompt new applications in mammalian cells.

Cas4

Cas4 is the type I system of class 1 and PD-(DE)Xk superfamily nuclease with three cysteines C-terminal cluster (Makarova et al., 2006) and possesses 5′ single-stranded DNA exonuclease activity (Zhang et al., 2012). The available structure is PDB:41C1.

Cas5

Cas5 is type of II large subunit cascade complex interacting family and Cas7 subunit and binding the 5′ handle of crRNA (Wiedenheft et al., 2011; Jore et al., 2011; Rouillon et al., 2013). In the subtype I-C system, Cas5 is the ribonuclease that has replaces Cas6 function (Koo et al., 2013). The available structures are PDB:3KG4; 3VZI;3VZH. There are two domains of RRM (ferredoxin) fold which is C-terminal domain is deteriorated in much Cas5 proteins of type I;RAMP Superfamily.

Cas6

Cas6 is directly involved in pre-crRNA transcript processing and metal-independent endoribonuclease that generates crRNAs (Hale et al., 2009; Niewoehner et al., 2014; Reeks et al., 2013; Richter et al., 2013; Carte et al., 2008; Haurwitz et al., 2010). The available structure is two domains of RRM (ferredoxin) fold, RAMP superfamily, PDB: 2XLJ, 1WJ9, 314H, 4C8Z and 4DZD.

Cas7

Cas7 is a type of I system and subunit of cascade complexes binding crRNA (Wiedenheft et al., 2011; Rouillon et al., 2013) which is often to present in Cascade and the structure PDB: 3ps0, 4N0L and RAMP superfamily.

Cas8

Cas8 has a subunit of cascade complex which involved in PAM recognition (Rouillon et al., 2013) and the available structure PDB:4AN8.

Cas9

Cas9 has a place with the class 2 sort II CRISPR frameworks and is the most broadly utilized genome altering enzyme. In particular, *Streptococcus pyogenes* Cas9 (SpCas9) was the first to be isolated outside from prokaryotic cells (Jinek et al., 2012) and reconstructed for genome altering in mammalian cells (Cong et al., 2013; Mali et al., 2013) it remains the most generally utilized Cas9. Subsequent DNA target acknowledgment, SpCas9 commonly produces a dull twofold strand break (DSB)13.

DNA focusing by SpCas9 depends on the 20-nucleotide long spacer and on the PAM 5′-NGG (N speaks to any nucleotide) (Jinek et al., 2012; Mojica et al., 2009) Cas9 enzymes are double RNA guided: a crRNA is liable for

DNA focusing on and furthermore hybridizes with the trans-activating crRNA, whichis applied for shaping the complex with Cas9 (Deltcheva et al., 2011; Gasiunas et al., 2012).

The PAM 5′-NGG limits the accessibility of SpCas9 target locales in the human genome to a normal of one objective site for each eight base pairs (Hsu et al., 2014). To expand the accessibility of target destinations, coordinated development approaches have created variations with changed PAM specificities (Kleinstiver et al., 2015; Hu et al., 2018) (Table 1). For instance, an extended PAM SpCas9 variation, xCas9, perceives 5′-NG, 5′-GAA and 5′-GAT PAM sequences (Hu et al., 2018).

Another inspiration for building Cas9 variations is to incrementally focus on specificity. A few studies have reported transformed Cas9 variations with diminished askew cleavage following articulation of Cas9 and gRNAs from plasmids (Slaymaker et al., 2016; Kleinstiver et al., 2016; Chen et al., 2017; Casini et al., 2018) or their transport as ribonucleoprotein complexes (Casini et al., 2018). Then again, on-target CRISPR-Cas particularity has been expanded by building optional structures as RNA clips on the spacer locale of gRNAs, which increment the thermodynamic hindrance to crRNA or gRNA strand intrusion at off-target destinations while by and large keeping up on-target activity (Vakulskas et al., 2018; Kocak et al., 2019).

Cas9 orthologues that perceive diverse PAM arrangements have given a more noteworthy decision of target locales. For instance, *Streptococcus thermophilus* Cas9 perceives the PAM 5′-NNAGAAW (W speaks to An or T) (Cong et al., 2013; Deveau et al., 2008) and *Neisseria meningitidis* Cas9 perceives 5′-NNNNGATT26-(Hou et al., 2013). These Cas9 orthologues have been used for DNA focusing in microorganisms and mammalian cells. Moreover, the PAM perceived by *Staphylococcus aureus* Cas9 (SaCas9) is 5′-NNGRRT (R connects to An or G) (Ran et al., 2015). Eminently, SaCas9 quality altering efficiencies are practically identical to those of SpCas9, and the littler size of SaCas9 (1,053 amino acids contrasted and 1,368 amino acids of SpCas9) has permitted its utilization in size-confined vectors, for example, adeno-related infection (AAV) (Ran et al., 2015).

All the more as of late, a considerably littler Cas9 orthologue, from *Campylobacter jejuni* (984 amino acids), was accounted for to perceive the PAM 5′-NNNVRYM (V speaks to A, C or G; Y speaks to C or T) 30 and utilized for focused genome altering *in vivo* (Yamada et al., 2017; Kim et al., 2017). Extra endeavors to distinguish Cas9 orthologues brought about the disclosure of CasX (980 amino acids), the little information has been given regarding Cas9 to date (Burstein et al., 2017).

Cas 10

It is a large subunit of cascade and Cmr complex form which having available structure PDB:3UNG, 4DOZ. They also form a homologous domain to the palm domain of nucleotide cyclases and also nucleic acid polymerases. There are two domains homologous to palm domain polymerases and cyclases, both belong to RRM (ferredoxin) fold link with Zn finger containing domain and C-terminal alpha-helical domain fusion: HD nuclease domain (Cocozaki et al., 2012).

Cas11

Cas11 is providing protection against mobile genetic elements but still, function is not known.

Cas 12a

Cas12a another class II RNA-guided endonuclease that has been reinvented for quality altering in human cells is Cas12a (once in the past Cpf1) (Zetsche et al., 2015). As a sort V framework, Cas12a produces an amazing cut with a 5′ overhang at DNA target locales and doesn't utilize a transactivating crRNA (Fig. 1b). Rather than the age of gruff finishes by Cas9, generation of amazing finishes by Cas12a might be favorable for applications, for example, coordinating DNA arrangements in an exact direction. Moreover, Cas12a can sever crRNA clusters to produce its own crRNAs. This crRNA handling capacity encourages the utilization of a solitary altered crRNA exhibit for disentangled multiplexed genome altering with various crRNAs34. Cas12a from *Acidaminococcus spp.* (AsCas12a) and Cas12a from *Lachnospiraceae spp.* (LbCas12a), the first Cas12a orthologues that were appeared to have movement in mammalian cells, perceive the PAM arrangement 5′-TTTV upstream of the objective grouping. To build its genome altering movement, an upgraded AsCas12a variation (enAsCas12a) has been engineered (Kleinstiner et al., 2019). To build the focus on the scope of Cas12a, AsCas12a variations were as of late built to perceive the PAMs 5′-TYCV and 5′-TATV36 or the PAMs 5′-VTTV, 5′-TTTT, 5′-TTCN and 5′-TATV35. The special highlights and cutting instrument of Cas12a give a genome altering apparatus that grows the CRISPR tool compartment.

CRISPR-Cas13

Cas13 has a place with the class 2 sort VI RNA-guided RNA nuclease and three proteins have been recognized; Cas13a (C2c2), Cas13b and Cas13c. Cas13a from *Leptotrichia wadei* (LwaCas13a) can be designed for focusing on endogenous transcripts in human HEK293FT cells and in plants. P5-125 (PspCas13b), which is C-terminally combined to the atomic fare signal (NES) area of the HIV Rev gene, has the most proficient RNA obstruction in mammalian cells (45, 46).

Mechanism of CRISPR-Cas system in Genome Editing

The editing of the gene with the help of CRISPR technologies are based on double-strand breaks (DSBs) in particular or specific regions of the genome and also repair own mechanism by cellular processes. The RNA guided DNA focused CRISPR/Cas technology was quickly adopted by researchers by the desirable quality of its affordability and easy to use in the field of genome editing as compared to earlier approaches for genome engineering such as transcription activator-effector nuclease (TALENs) and zinc-finger nuclease (ZFNs). CRISPR/Cas approaches have their own origin in an immune defense mechanism found in bacteria which provides immunity to the host against nucleic acids such as phages and viruses (Barrangou et al., 2007). There are three types of CRISPR/Cas approaches in each several subgroups according to the most famous classification (Makarova et al., 2011). Type II is the most commonly used system for gene editing and it also consists of three components such as transactivating crRNA (tracrRNA), CRISPR RNA (crRNA) and endonuclease (Cas9) (Jinek et al., 2012). The tracrRNA and crRNA both formed duplex structures is known as guide RNA (gRNA) which can be replaced by a chimeric single (sgRNA) (Jinek et al., 2012). The 20 base–pair (bp) sequences in sgRNA which is designed to be complementary to the target site and also must be followed by the PAM sequences called as "protospacer-adjacent motif", which is required for compatibility with Cas9 Protein.

The ribonucleoprotein (RNP) is formed when the sgRNA and Cas9 nuclease are expressed in the cell and guided by sgRNA to target the DNA site. The DNA is cleaved to produce a DSB when sgRNA binds to the target sequence by Cas9 and Watson-Crick base-pairing. The generating blunt ends when cleavage occurs within the protospacer (within three nucleotides upstream the PAM). The cleavage of opposite DNA strands positive (+) and negative (-) due to RuvC and HNH protein involvement on active site of Cas9 (Gasiunas et al., 2012; Mojica et al., 2009).

The homology-directed repair (HDR) and non-homologous end joining (NHEJ) are repairs DSB in cellular machinery which is basically depending on the presence of a repair template and also by the cell condition (Liang et al., 1998; Bibikova et al., 2001; Fishman-Lobell et al., 1992). The donor DNA template is used by the HDR pathway which is recombined at the site of DSB and repair in an accurate condition. The explanation about specific sequences or changes occurs in the target region of the genome by homology-directed repair. The randomly added or deleted nucleotides at the DSB site region by the NHEJ pathway is the wrong system so therefore it can be used or induce for specific gene knockouts (KO) by frameshift mutations. The gene can edit by multiple loci at the same time by sgRNA targeting sites and also with the help of the CRISPR/Cas9 system (Jakociunas et al., 2015; Wang et al., 2013; Li, W et al., 2013). The small deletions in the same cell by using two sgRNA and form complex rearrangements (Torres et al., 2014; Choi et al., 2014) even the whole chromosome at suppression condition (Adikusuma et al., 2017). The CRISPR/Cas9 is customizations of CRISPR/Cas9 components with different interactors which improved the specificity of the system and exploited their applications at the stage of editing (Dominguez et al., 2016). There are so many studies that have established by the presence of off-target activity so improving DNA specificity for CRISPR technology field (Tsai et al., 2016). There are several strategies developed to decrease the ratio of off-target products. The specificity of CRISPR/Cas9 needs to be further improved and development of an effective technology for gene editing.

CRISPR-Cas System in Drug Discovery

The development of new drug discovery is a complex process for the identification of new drugs and also bringing them to market for sale. This is totally based on the hypothesis by a biologist for a particular target effect on changes of a specific disease by designing a drug. The condition for identity of a specific gene to make a drug with the help of genome engineering which is responsible for causing a disease. This is time-consuming process. So, we focus on CRISPR/Cas system engineering which is a good or positive effect on drug discovery by target molecules whose involve for inhibition and activation causes to prevent disease (Ahmad et al., 2018). The cells survive under particular conditions for drug treatment through identify genes with positive selection. For example, cells can be treated with CRISPR sequences and then exposed to an anti-cancer drug. Only drug-resistant survivors can be harvested to analyze the sequence of the gRNAs, which are used to identify candidate genes for drug resistance (Kurata et al., 2018). The slow-growing cells or dead cells are efficiently used under a particular condition by negative selection. This is helpful to identify particular genes which are essential for survival and can be promising candidates for targeted drugs.

CRISPR/Cas9 System in Drug Resistance

Drug discovery is the identification of genes involved in drug resistance which is an important application of CRISPR/Cas9 technology. The anticancer drug-resistant agents are evaluated by the process of mutagenesis across the cell division. Only those cells survive which carrying mutation that impairs the action of drug application (Guichard et al., 2018). We also screen particular gene deletions with drug resistance through the CRISPR/Cas9 system. Cells are acquiring resistance to the specific drug which exposed to a pool of CRISPR/Cas9 gRNAs that target various genes for knocked out. Cells become sensitive to drug exposure which is confirmed by drug resistance. Selected genes identified as resistant to the drug then be targeted with different or other drugs to avoid the emergence of resistance (Scott et al., 2018). For example, HPRT1 gene disrupts by CRISPR/Cas9 editing which is resistant to 6-thioguanine (conventional anti-cancer drug) (Smurnyy et al., 2014), Equally, the XPO1 gene mediated by CRISPR/Cas9 system confers resistance to selinexor (Neggers et al., 2015).

FUTURE APPLICATIONS TO TARGETING HOST AND VIRAL GENES

CRISPR technology has been developed and applied to antiviral research in animals and humans. Based on virus-host interactions and the fundamental rules of nucleic acid cleavage or gene activation by the CRISPR-Cas framework, it tends to be utilized to target both the host and viral factor to clear popular repositories and virus infection or replication. Here, we outline the later progress of the CRISPR-Cas innovation in altering host genes as an antiviral system. Virus host interactions are dynamic procedures. The virus misuses the host factor to complete its life cycle and the host utilizes an immune defense system to inhibit viral infection. Notwithstanding the natural insusceptible framework and versatile resistant mechanical assembly, have cells been developed to express broad limitation factors that level infection replication. To check this, infections utilize explicit systems to escape insusceptible observation. Hence, CRISPR-Cas frameworks to treat virus which can be directed by focusing on viral nucleic acid, just as host factors including positive and negative controllers. We look at the changed CRISPR-Cas system and its application in antiviral research by focusing on host factors. We likewise examine the latest research in different CRISPR-Cas systems in focusing on the host factor. Host dependent factor targeted by CRISPR-Cas systems needs to be summarized in future applications.

BASICS OF GENETIC ENGINEERING

The DNA Toolbox

Genetic Engineering and Its Applications

- Most methods for cloning pieces of DNA in the laboratory share general features, such as the use of bacteria and their plasmids.
- Plasmids are small circular DNA molecules that replicate separately from the bacterial chromosome.
- Cloned genes are useful for making copies of a particular gene and producing a protein product.

Construction of Recombinant DNA Using Plasmid pUC19

Genetic Engineering and Its Applications

- Gene cloning involves using bacteria to make multiple copies of a gene.
- Foreign DNA is inserted into a plasmid, and the recombinant plasma is inserted into a bacteria cell.
- Reproducing in the bacteria cell results in the cloning of the plasmid including the foreign DNA.
- This results in the production of multiple copies of a single gene.

Using Restriction Enzymes to Make Recombinant DNA

- Bacterial restriction enzymes cut DNA molecules at specific DNA sequences called restriction sites.
- A restriction enzyme usually makes many cuts, yielding restriction fragments.

- The most useful restriction enzymes cut DNA in a staggered way, producing various fragments with "sticky ends" that bond with complementary sticky ends of other fragments.

Types of Restriction Endonucleases

Among these restriction enzymes, type II enzymes are generally used in genetic engineering experiments. All the restriction enzymes currently available on the market are type II.

Transformation Procedure

Screening a Library for Clones Carrying a Gene of Interest

- A clone carrying the gene of interest can be identified with a nucleic acid probe having a sequence complementary to the gene.
- The process is called nucleic acid hybridization.

Screening a Library for Clones Carrying a Gene of Interest

Eukaryotic Chromosome Organization

- Eukaryotic chromosomes are long, linear DNA molecules.
- Three types of DNA sequences are required for chromosome replication and segregation.
- Origins of replication (multiple)
- Centromeres (1)
- Telomeres (2)

Eukaryotic Chromatin Compaction

- If stretched end to end, a single set of human chromosomes will be over 1 meter long.
- Yet the cell's nucleus is only 2 to 4 um in diameter.
- Therefore, the DNA must be tightly compacted to fit.
- The compaction of linear DNA in eukaryotic chromosomes involves interactions between DNA and various proteins.
- Proteins bound to DNA are subject to change during the life of the cell.
- These changes affect the degree of chromatin compaction.

Nucleosomes

- The repeating structural unit within eukaryotic chromatin is the nucleosome.
- It is composed of double- stranded DNA wrapped around an octamer of histone proteins.
- An octamer is composed of two copies of four different histones.
- 146 bp of DNA make 1.65 negative superhelical turns around the octamer.
- The overall structure of connected nucleosomes resembles "beads on a string".
- This structure shortens the DNA length about seven-fold.

Nucleosomes Join to Form a 30 nm Fiber

- Nucleosomes associated with each other to form a more compact structure termed the 30 nm fiber
- Histone H1 plays a role in the compaction.
- At moderate salt concentrations, H1 is removed.
- The result is the classic beads-on-a-string morphology.
- Beads associate together into a more compact morphology.

Determining Gene Function

- Gene expression can also be silenced using RNA interference (RNAi).
- Synthetic double-stranded RNA molecules matching the sequence of a particular gene are used to break down or block the gene's mRNA.

Human Gene Therapy

- Gene therapy is the alteration of an afflicted individual's genes.
- Gene therapy holds great potential for treating disorders traceable to a single defective gene.
- Vectors are used for the delivery of genes into specific types of cells. For example bone marrow.

- Gene therapy raises ethical questions, such as whether human germ-line cells should be treated to correct the defect in future generations

DNA Repair

- Replication itself can very occasionally damage the information content in DNA when errors introduce mismatched base pairs (such as G paired with T).
- The chemistry of DNA damage is diverse and complex.
- Cells require a means for the repair of missing, altered or incorrect bases, bulges due to insertion or deletion, UV-induced, pyrimidine dimmers, strand breaks or cross-links.
- The cellular response to this damage includes a wide range of enzymatic systems that catalyze some of the most interesting chemical transformations in DNA metabolism.

Consequences of DNA damage

- Change in the base sequence of the DNA, which, if replicated and transmitted to future cell generations, becomes permanent-Mutation.
- Silent mutation-

a) negligible effect on the function of a gene

- Can lead to cancer

Type of DNA Repair

- Mismatch repair
- Base-excision repair
- Nucleotide-excision repair
- Direct repair
- Recombination repair

Enzymes involved in DNA Repair

Enzymes/proteins	Type of
damage	
Mismatch repair	
Dam methylase	
MutH, MutL,MutS proteins	
DNA helicase II	
SSB	
DNA polymerase III	Mismatches
Exonuclease I	
Exonuclease IV	
RecJ nuclease	
Exonuclease X	
DNA ligase	
Basse-excision repair	
DNA glycosylases	Abnormal bases
(uracil, hypoxanthine,xanthine);	
AP endonucleases	alkylated bases; in
some pyrimidine dimers	
DNA polymerase I	
DNA ligase	
Nucleotide-excision repair	
ABC excinuclease	DNA lesions that
cause large structural changes	
DNA polymerase I	(e.g., pyrimidine
dimmers)	
DNA ligase	
Direct repair	
DNA photolyases	pyrimidine dimmers
O^6-Methylguanine	
O^6 –Methylguanine-DNA	1-Methylguanine,
3-Methylcytosine	
Methyltransferas	
AlkB protein	

Double-Strand break Repair

Recombination Repair

Nonhomologous End Joining (NHEJ)

Recombination Repair

NHEJ

- The broken ends of the DNA must be aligned, its frayed ends trimmed and/or filed in, their strands ligated.
- Core NHEJ machinery in eukaryotes includes the DNA end-blinding protein Ku (Ku70 and Ku80), DNA ligase IV, and the accessory protein Xrcc4.

What is genome editing?

- A genetic engineering approach in which DNA is inserted, removed or replaced at a precise location within the gnomic.
- Engineered nucleases.
- Recombination-based approaches Viral genomic editing: rAAV:

Nuclease based genomic editing:

The core technologies now most commonly used facilitate genome editing are:

- Clustered regularly interspaced short palindromic repeats (CRISPR)-CRISPR-associated protein 9 (Cas9)
- Transcription activator-like effector nucleases (TALENs),
- Zinc-finger nucleases (ZFNs)
- Homing endonucleases or meganucleasses.
    - Creation of isogenic cell lines -> only differ by the change we've introduced.
    - Make (nearly) any modification we desire

Zinc – Finger nucleases

- ZFNs, which are fusions between a custom-designed Cys2-His2 zinc-finger protein and the cleavage domain of the Fok1 restriction endonuclease, was the first targeted nucleases to achieve widespread use.
- ZFNs, function as dimers, with each monomer recognizing a specific "half -site" sequence—typically nine to 18 base pairs (bps) of DNA-via the zinc-finger DNA binding domain.
- Dimerization of the ZFN proteins is medicated by the Fok1

clcavage domain, which cuts DNA within a five- to seven-bp spacer sequence that separates two flanking zinc-finger binding sites.

- Each ZFN is typically composed of three or four zinc-finger domain, with each individual domain composed of ~30 amino acid residues that are organized in a ββχ motif.

Zinc-Finger Nucleases

The residues that facilitate DNA recognition are located within the ϰ-helical domain and typically interact with three bps of DNA, with occasional overlap from an adjacent domain.

Using methods such as phage display, a large number of zinc-finger domains recognizing distinct DNA triplets have been identified.

One major concern associated with the use of ZFNs for genome editing (in addition to all targeted nucleases) is off-target mutations.

In particular, it remains challenging to create zinc-finger domains that can effectively recognize all DNA triplets, especially those of the 5′-CNN-3and 5″-TTNN-3″ variety.

TALE Nucleases

TALE proteins are bacterial effectors. In2009, the code used by TALE proteins to recognize DNA was uncovered.

In a matter of months, this discovery enabled the creation of custom TALENs capable of modifying nearly any gene.

Like ZFNs, TALENs, are modular in form and function, comprised of an amino-terminal TALE DNA-binding domain fused to a carboxy-terminal foci cleavage domain, which cuts within a 12- to 19-bp spacer sequence the separates each TALE binding site.

TALEs are typically assembled to recognize between 12-to 20-bps of DNA, with more base typically leading to higher genome-editing specificity.

TALE Nucleases

The TALE-binding consists of a series of repeat domains, each ~34 residues in length. Each repeat contacts DNA via the amino acid residues at positions 12 and 13, known as the repeat variable diresidues (RVDs).

Unlike zinc fingers, which recognize DNA triples, each TALE repeat recognizes only a single bp, with little to no target site overlap from adjacent domains.

The most commonly used RVDs for assembling synthetic TALE arrays are NI for adenine, HD for cytosine, NG for thymine, and HN for guanine or adenine.

Compared to ZFNS, TALENs offer two distinct advantages for genome editing.

TALE Nucleases

First, no selection or directed evolution is necessary to engineer TALE arrays, dramatically reducing the amount of time and experience needed to assemble a functional nuclease.

Second, TALENs have been reported to show improved specificity and reduced toxicity compared to some ZFNs, potentially because of their increased affinity for target DNA or perhaps a greater energetic penalty for associating with base mismatches.

What is CRISPR-Cas9?

CRISPR technology was adapted from the natural defence mechanisms of bacteria and archaea (the domain of single-celled microorganisms).

These organisms use CRISPR-derives RNA and various Cas proteins, including Cas9, to foil attacks by viruses and other foreign bodies.

They do so primarily by chopping up and destroying the DNA of a foreign invader. When these components are transferred into other, more complex organisms, it allows for the manipulation of genes, or "editing".

CRISPR-Cas9: the key players

CRISPRs: "CRISPR" stands for "clusters of regularly interspaced short palindromic repeats".

It is a specialized region of DNA with two district characteristics: the presence of nucleotide repeats and spacers.

Repeated sequences of nucleotides – the building blocks of DNA –are distributed throughout a CRISPR region. Spacers are bits of DNA that are interspersed among these repeated sequences.

In the case of bacteria, the spacers are taken from viruses that previously attacked the organism. They serve as a bank of memories, which enable bacteria to recognize the viruses and fight off future attacks.

CRISPR –Cas 9: The key players

This was first demonstrated experimentally by Rodolphe Barrangou and a team of researchers at Danisco, a food ingredients company.

In a 2007 paper published in the journal Science, the researchers used *Streptococcus thermophilus* bacteria, which are commonly found in yogurt and other dairy cultures, as their model._They observed that after a virus attack, new spacers were incorporated into the CRISPR region.

Moreover, the DNA sequences of these spacers were identical to parts of the virus genome. They also manipulated the spacers by taking them out or putting in new viral DNA sequences.

In this way, they were able to alter the bacteria's resistance to an attack by a specific virus. Thus, the researchers confirmed that CRISPRs play a role in regulating bacterial immunity.

CRISPR –Cas 9: The key players

CRISPR RNA (crRNA)

Once a spacer is incorporated and the virus attacks, a portion of the CRISPR is transcribed and processed into CRISPR RNA, or "crRNA".

The nucleotide sequence of the CRISPR acts as a template to produce a complementary sequence of single-stranded RNA.

Each crRNA consists of a nucleotide repeat and a spacer portion, according to a 2014 review by Jennifer Doudna and Emmanuelle Charpentier.

CRISPR –Cas 9: The key players

Cas9: The Cas9 protein is an enzyme that cuts foreign DNA.

The protein typically binds to two RNA molecules:

crRNA and another called tracrRNA (or "trans-activating crRNA").

The two then guide Cas9 to the target site where it will make its cut. This expanse of DNA is complementary to a 20-nucleotide stretch of the crRNA.

Using two separate regions, or" domains" on its structure, Cas9 cuts both strands of the DNA double helix, making what is known as a "double-stranded break,".

CRISPR –Cas9: The key players

There is a built-in safety mechanism, which ensures that Cas9 doesn't just cut anywhere in a genome.

Short DNA sequences known as PAMs ("protospacer adjacent motifs") serve as tags and sit adjacent to the target DNA sequence.

If the Cas9 complex doesn't see a PAM next to its target DNA sequence, it won't cut this is one possible reason that Cas9 doesn't ever attack the CRISPR region in bacteria.

An overview of CRISPR and NHEJ/HDR.

Genome Engineering Applications

Genome editing with CRISPR is not only simple, but also scalable.

Multiple gRNAs targeting multiple loci can be easily expressed in the same cell or organism.

The ability to precisely target CRISPR to a given locus makes it especially amenable to genetic screens.

Pooled gRNA libraries can be used to identify genes that are important to a given phenotype.

Current libraries are available for gene knockout, as well as transcriptional activation or repression.

Combined with the power of next-generation sequencing, CRISPR is a robust system for genome-wide screening.

Day 3 Screening for new drug targets And discovery workflow

How are drugs discovered and developed?

Introduction

- Drug: Compounds used for the prevention and treatment of diseases like cancer, etc.

- Ideal drug:
  - Target: bio-molecule, involved in signaling or metabolic pathways, that are specific to the disease process by either protein-protein or protein-nucleic acid interactions.
  - Antagonist action-inhibiting functions of the disease-causing proteins.
  - Inhibiting interactions of the proteins.
  - Activates other proteins that are deregulated in such diseases like cancer.

The Pharma Value Chain

Drug Discovery Animal Studies Clinical Tests Commercialization

The Pharma Value Chain

Building a library of gene protein (genome/proteome) sequences to mine for information

The Pharma Value Chain

- Look for protein or mRNA expressed (or not expressed) in a disease. Comparative gene expression assays, Comparative proteomic profiles.
- Look for genes/proteins essentials for an infectious agent and distinct from host genes/proteins.
- Look for gene and gene modifications associated with a disease.
- Find regulatory pathways required for the disease process.

The Pharma Value Chain

- Molecular level
- Screen enzyme inhibitors or activators.
- Cellular level
- Verify the involvement of the protein in the disease state (often use gene silencing siRNAs)
- Understand the protein pathway and interactions.

- Verify the critical nature of target and uniqueness.

The Pharma Value Chain

- Discover leads that affect the target gene, protein or pathway
- Inhibit defective protein
- Activate a defective protein.
- Inhibit expression of a protein/pathway.
- Activate expression of required protein/pathway.
- Stimulate protein modification or cellular location.

The Pharma Value Chain

- Evaluate leads to 'cure' the problem, e.g.: Replace missing or defective protein with gene therapy.
- Anti-sense or siRNA to prevent protein expression.
- Antibody to remove or inhibit protein target
- Stimulation of synthesis to replace or activate protein
- Stimulate protein modification or location

Drug Discovery Methods

- Screening natural compound collection

Natural Compound Library Screening

Drug Discovery Methods

- Screening natural compound collections
- Screening corporate compound collections
- *In silico* screening (Autodock)

In silico screening with Autodock

Drug Discovery Methods

- Screening natural compound collections
- Screening corporate compound collections

- *In silico* screening (Autodock)
- Rational drug design

Rational Drug Design for HIV Protease

Drug Discovery Methods

- Screening natural compound collections
- Screening corporate compound collections
- *In silico* screening (Autodock)
- Rational drug design
- Combinatorial chemistry

Combinatorial Chemistry

Drug Discovery Methods

- Lead Discovery
    - Screening natural compound collections
    - Screening corporate compound collections
    - *In silico* screening (Autodock)
    - Rational drug design
    - Combinatorial chemistry
- Lead validation
- Lead optimization

ADMET: Ideal Properties Of Drugs

- Absorption –Passes GI tract into the bloodstream
- Distribution – Gets to target tissue (blood-brain barrier)
- Metabolism – Not readily secreted
- Excretion – Not readily secreted.
- Toxicity – Not toxic to other cells or tissues.

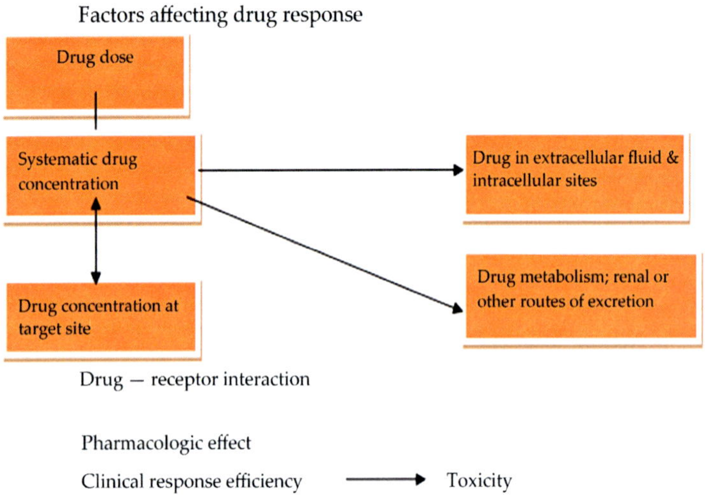

Chris Lipinski's Rule of Five

Lipinski and his Pfizer co-workers looked over a data set of drug candidates and noticed that there were some reasonably clear cutoffs for oral absorption and general cell permeability. They suggested that you need:

- Fewer than five hydrogen bond donors (which can be estimated by counting the total number of OH and NH groups in the molecule).
- Fewer than 5 hydrogen-bond acceptors (estimated by the total of N and O atoms in the molecule)
- A molecular weight of less than 500
- A partitioning coefficient (logP) of less than 5

The "rule of five" name came from the cutoffs all being multiple of five, in case you are wondering why there are only four rules.

The Pharma Value Chain

- Animal tests of toxicity and efficacy of therapy
  - Rodents (mice and rats)
  - Mammals (pigs)
  - Primates (monkey and chimpanzees)
  - Mouse lemurs (Microcebus)

The New Primate: Mouse Lemurs

(Microcebus margotmarshae)

The Pharma Value Chain

A small group of healthy volunteers (10's) to determine safety and toxicity. Maybe some members of the target group

The Pharma Value Chain

100's of the patient population to determine efficacy, dosage, safety.

The Pharma Value Chain

1000's of patients and control (normals) to determine efficacy, dosage, safety, side effects and interactions. Each prospective patient group (men, women, children, elderly and ethnic groups)

Drug designing is:

- Challenging
- Expensive
- Time-consuming

So, Multidisciplinary approach:

Computational tools, methodologies for structure-guided approach + Global gene expression data analysis by software.

Hence,

- Efficiency increased
- Cost-effectiveness
- Time saved
- Strategies to overcome toxic side effects

Drug Design

2 ways:

Development of ligands with desired properties for targets having known structure and functions.

Development of ligands with predefined properties for targets whose structural information may be or may not be known.

This, unknown target information can be found by global gene expression data.

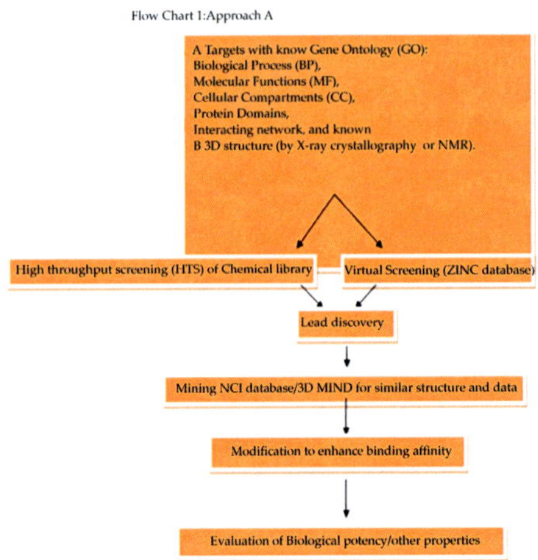

Figure 1. Possible steps in drug designing for known target.

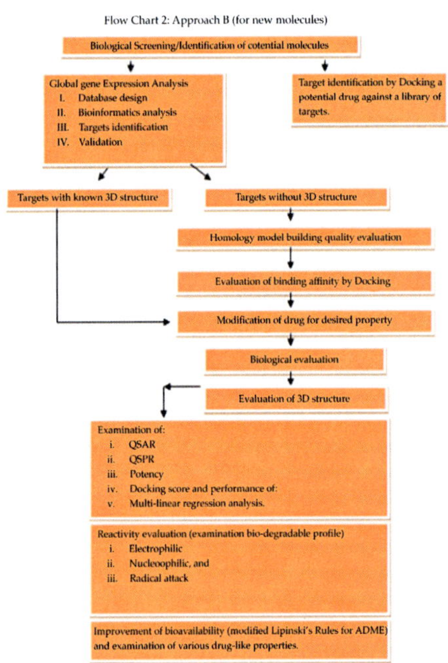

Figure 2

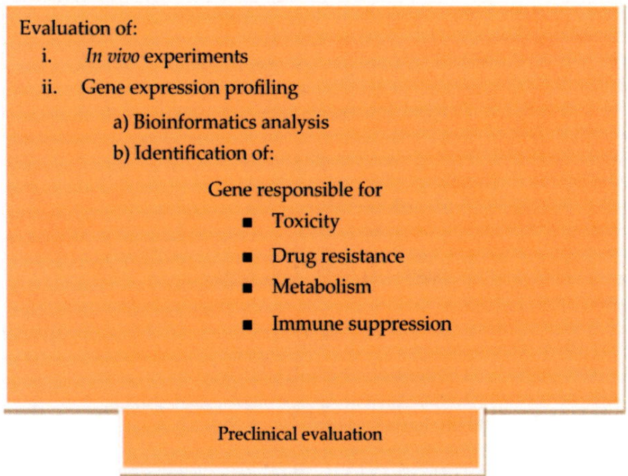

Figure 3: Additional properties towards the improvement of drug-like properties.

DRUG DISCOVERY, DELIVERY AND DEVELOPMENT

Target Identification & Validation in Drug Discovery

Outline

- The key to good drug design is working out and capturing the clinical spectrum of diseases and the exact role a potential therapeutic target plays in the disease.
- Discussion points
  - The properties of an attractive drug target
  - Outlines the approaches used to identify target
  - The key steps involved in target validation and
  - The benefits and challenges of using siRNA to validate targets

Target Identification

- Identifying a biological target that is 'druggable'-
  - A target is termed 'druggable' if its activity (behavior or

function) can be modulated by a therapeutic-whether it be a small molecule drug, or biologic.
  - Proteins and nucleic acids are both examples of biological targets.
- But what makes a 'good' target?

Properties of a Promising Drug Target

- The target has a confirmed role in the pathophysiology of a disease and/or is disease-modifying.
- Target expression is not evenly distributed throughout the body.
- The target's 3D-structure is available to assess druggability.
- The target is easily 'assayable' enabling high-throughput screening.
- The target possesses a promising toxicity profile; potential adverse effects can be predicted using phenotypic data.
- The proposed target has a favorable intellectual property (IP) status.(relevant for pharma companies)

Strategies for Identifying Drug Targets

- Many targets are initially identified using scientific literature and public databases such as DrugBank or TTD.
- This popular method for identifying drug targets can result in competition amongst drug developers.
- When it comes to advancing the drug discovery process some are willing to accept the more scientific risk to drive forward targets that perhaps have less information available, but also less competition, whereas others are bolder and are willing to accept greater competition in the hopes of driving 'more established' targets forward.

Strategies for Identifying Drug Targets

Besides the literature, there are numerous other target identification strategies, which can be based on one of two target identification strategies.

The first is target deconvolution-in this scenario you begin with a drug that appears efficacious and the target is identified retrospectively.

The other is known as target discovery-this approach works on the concept that 'if you want a new drug you must find a new target', once a target is identified compound libraries are screened to find a drug that binds to the target and elicits the desired effect.

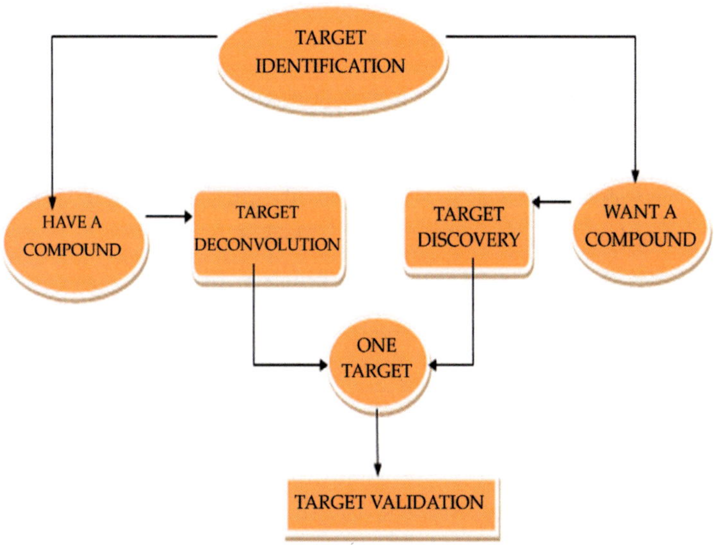

Target Deconvolution vs Target Discovery

The phenotypic approach to drug discovery falls within the realm of target deconvolution and involves exposing cells, isolated tissues, or animal models, to small molecules to determine whether a specific candidate molecule exerts the desired effect- which is observed by a change in phenotype.

Whilst numerous animal models can be used for the characterization of small molecules and small-scale drug screening approaches, the use of mammalian cells is often favored due to their compatibility with high-throughput screening (HTS) and greater physiological relevance.

The phenotypic approach goes beyond individual proteins or nucleic acids and involves the study of entire signaling pathways. The drug's effect is determined before the specific biological (drug) target that underlies the observed phenotypic response is identified.

Target de-convolution

- Target de-convolution can be achieved by numerous methods including;

- Affinity chromatography,
- Expression – cloning
- Protein microarray
- ‘Reverse transfected’ cell microarray, and
- Biochemical suppression.

Advantages and Challenges of Phenotypic Drug Discovery

- The greatest advantages phenotypic approaches have over target-based are their ability to demonstrate the efficiency of a drug in the context of a cellular environment.
- The drug is acting on the target in its ‘normal biological context’, rather than on a purified target in a biochemical screen.
- Cost, availability of cells, complex assay methodology, and throughput are all potential challenges associated with cell-based phenotypic screens.
- However, as assays become miniaturized and the use of three-dimensional cell models (organoids and spheroids) continues to gain momentum, both scalability and physiological relevance have been improved, leading to greater adoption of phenotypic approaches.

Target Discovery

- In target-based drug discovery, biological (drug) targets are already established (or ‘discovered’) before lead discovery starts-hence target discovery is the cornerstone of target-based screening.
- The target’s role in a disease process is known, this target is then used to create relevant systems-based assays, and vast compound libraries are screened in search of a ‘hit’- a candidate drug.

Advantages and Challenges of Target-based Discovery

Target-based drug discovery can exploit numerous approaches (including crystallography, computational modeling, genomics, biochemistry, and binding kinetics) to uncover exactly how a drug interacts with the target of interest, enabling:

- Development of the structure-activity relationship (SAR) (the relationship between the structure and biological activity of a molecule)
- Development of biomarkers:

Techniques	Drug discovery approach
Affinity chromatography	Target deconvolution
Expression-cloning	Target deconvolution
Protein microarray	Target deconvolution
Reverse transfected cell microarray	Target deconvolution
Biochemical suppression	Target deconvolution
Sirna	Target deconvolution discovery/ validation
DNA microarray	Target discovery
Systems biology	Target discovery
Study of existing drugs	Target discovery

Discovery of future therapeutics that act at the specific target of interest

An overview of target deconvolution, discovery and validation techniques

Validating Drug Targets

- Target validation is the process of demonstrating the functional role of the identified target in the disease phenotype.
- Whilst the validation of a drug's efficacy and toxicity in numerous disease-relevant cell models and animal models is extremely valuable-the ultimate test is whether the drug works in a clinical setting.

Validating Drug Targets

- Target validation can be broken down into two key steps.
- Reproducibility. Once a drug target is identified, whether it be via a specific technique or from the review of literature, the first step is to repeat the experiment to confirm that it can be successfully reproduced.
- Introduce variation to the ligand (drug) –target –environment.
  - It should be possible to modulate the drug's affinity to the

target by modulating the activity of the drug molecule.

- Verifying the cell or tissue type, should or should not, alter the drug's effect.
- Introducing mutations into the binding domain of the protein target should result in either modulation or loss of activity of the ligand (drug).

siRNA for Target Validation

- Several different target-validation techniques can be used to elucidate target function, however, by far the most popular and most widely used approach utilizes small interfering RNAs (siRNAs).
- siRNA allows you to mimic the effect of the drug via mRNA modulation resulting in the temporary suppression of a gene-product-your drug target.
- It is therefore possible to demonstrate the 'value' of the target without actually using/having the drug.
- By observing the phenotypic effect that results from a decrease in the target protein you can confirm whether the target warrants further development.

Advantages and Disadvantages of siRNA

Advantages	**Disadvantages**
You can investigate the effect of inhibiting a potential target without having a drug	Down-regulating a gene is not the same as inhibiting a specific region of the gene-product (target)
siRNA enables to you mimic the effect of a drug more accurately than if you were to 'knockout' a gene.	Down-regulating a gene can have a more exaggerated effect compared to that seen from decreasing the concentration of the gene-product (target)
Knowledge of protein structure is not required to use siRNA	There may still be some protein synthesis occurring as it is not possible to achieve 100% down-regulation with siRNA
siRNA is an inexpensive method of target validation	siRNA delivery is challenging.

Day 5 HIT and Lead Optimization and CRISPR Tailored Disease Models

How CRISPR Is Accelerating Drug Discovery

- The drug discovery process, in which compounds are evaluated for therapeutic use, has resulted in safe and effective therapies for a variety of diseases. However, assessment of new compounds for drug development are notoriously long and costly: they typically span more than a decade and exceed a billion dollars. Furthermore, only a small percentage of candidates actually make it to market.
- High costly make the development of new drugs a risky endeavor for pharmaceutical companies and consequently may hinder the exploration of new therapeutics. Removing these barriers using new technologies like CRISPR is key to expanding drug discovery.

How CRISPR Is Accelerating Drug Discovery

Hit to Lead (H2L) Process in Drug Discovery

As the name suggests, the hit ti lead (H2L) stage of drug discovery aims to discover compounds of interest (leads) from a selection of compounds (hit) that demonstrate promising therapeutic effects. The hit to lead the process, which is also referred to as lead generation, typically occurs following the completion of target validation, assay development, and high-throughput screening studies.

What is a hit?

- Hits provide researchers with important information on the relationship that exists between the molecular structure, identity, and purity of the compound. This plays a significant role in its biological activity with the target.
- The main approach to hit discovery is high throughput screening (HTS); assays that are used to screen an entire compound library against a new drug's target. Knowledge-based screening (also known as focused-based screening), is another crucial aspect of hit discovery. This involves a thorough analysis of a chemical library

consisting of smaller subsets of molecules with potential activity at the target site. Fragment-based screening and physiological screening processes may also be used to complement these approaches during the drug discovery process.

What is a hit?

Once a series of hits have been discovered, researchers have a wide range of assays they can utilize for the hit confirmation process, all of which vary depending on the nature of the drug's target, the compound screening scale, and available equipment. Pharmacological assays, for example, are used to identify compounds that exhibit the desired mechanism of action. Assay reproducibility, quality and sensitivity are particularly important during this stage.

Challenges of Hit to Lead (H2L)

- Some of the most significant aspects hit to lead involve assessing the pharmacokinetics and toxicological parameters of a new drug, as well as optimizing its potency.
- One of the most common challenges that medicinal chemists face during a hit to lead is determining how to optimize one property (such as absorption) without compromising another property (such as the drug's high potency). For example, since the reduction in a compound's molecular weight can exhibit negative impacts on its overall potency.
- To overcome these challenges, parallel optimization of potency, selectivity, pharmacokinetics and physical properties is often conducted. This multiparametric approach collects data on each of the aforementioned drug properties. In doing so, researchers are able to develop drugs that exhibit more uniform characteristics and that will provide more information to their fate during later pre-clinical and clinical studies.

Hit to lead

- Hit to lead (H2L) also known as lead generation is a stage in early drug discovery where small molecule hits from a high throughput screen (HTS) are evaluated and undergo limited optimization to identify promising lead compounds.[1][2] These lead compounds undergo more optimization in a subsequent step of drug discovery called lead optimization (LO).[3][4] The drug discovery process generally follows the following oath that includes a hit to lead stage:

- Target Validation (TV) →Assay Development →High-Throughput Screening (HTS) →Hit to Lead (H2L) →Lead Optimization (LO) →Pre-Clinical Drug Development →Clinical Drug Development.
- The hit to lead stage starts with confirmation and evaluation of the initial screening hits and is followed by a synthesis of analogs (hit expansion). Typically the initial screening hits display binding affinities for their biological target in the micromolar (10^{-6} molar concentration) range. Through limited H2L optimization, the affinities of the hits are often improved by several orders of magnitude to the nanomolar (10^{-9}M) range. The hits also undergo limited optimization to improve metabolic half-life so that the compounds can be tested in animal models of disease and also to improve selectivity against another biological target binding that may result in undesirable side effects.
- On average, only one in every 5,000 compounds that enter drug discovery to the stage of pre-clinical development becomes an approved drug.

Hit Confirmation

After hits are identified from a high throughput screen, the hits are confirmed and evaluated using the following methods:

- Confirmatory testing: compounds that were found active against the selected target are re-tested using the same assay conditions used during the HTS to make sure that the activity is reproducible
- Dose -response curve: the compound is tested over a range of concentrations to determine the concentration that results in half-maximal binding or activity (IC50 or EC50 value respectively).
- Orthogonal testing: confirmed hits are assayed using a different assay that is using closer to the target physiological condition or using a different technology.
- Secondary screening: confirmed hits are tested in a functional cellular assay to determine efficacy.

- Synthetic tractability: medicinal chemists evaluate compounds according to their synthesis feasibility and other parameters such as up-scaling or cost of goods.
- Biophysical testing: nuclear magnetic resonance (NMR), isothermal titration calorimetry (ITC), dynamic light scattering (DLS), surface Plasmon resonance (SPR), dual-polarization interferometry (DPI), microscale thermophoresis (MST) are commonly used to assess whether the compound binds effectively to the target, the kinetics, thermodynamics, and stoichiometry of binding, any associated conformational change and to rule out promiscuous binding.
- Hit ranking and clustering: confirmed hit compounds are then ranked according to the various hit confirmation experiments.
- Freedom to operate evaluation: hit structures are checked in specialized databases to determine if they are patentable.

Hit Expansion

Following hit confirmation, several compounds will be chosen according to the previously define tests.

An ideal compound cluster will contain members that possess:

- High affinity towards the target (less than 1 µM)
- Selectivity versus other targets
- Significant efficacy in a cellular assay
- Drug likeness (moderate molecular weight and lipophilicity usually estimated as ClogP). Affinity, molecular weight and lipophilicity can be linked in single parameters such as ligand efficiency and lipophilic efficiency.
- Low to moderate binding to human serum albumin
- Low interference with P450 enzymes and P-glycoproteins
- Low cytotoxicity
- Metabolic stability
- High cell membrane permeability
- High water solubility (above 10 µM)
- Chemical stability
- Synthetic tractability
- Patentability

The project team will usually select between three and six compound series to be further explored. The next step will allow the testing of analogous compounds to determine a quantitative structure-activity relationship (QSAR). Analogs can be quickly selected from an internal library or purchased from commercially available sources ("SAR by catalog " or "SAR by purchase"). Medicinal chemists will also start synthesizing related compounds using different methods such as combinatorial chemistry, high-throughput chemistry, or more classical organic chemistry synthesis.

Hit Validation

- Hits identified through the initial screens are validated through a variety of cell-based assays. CRISPR has profoundly impacted this stage of preclinical development by promoting the generation of models that accurately recapitulate diseases.
- Instead of being limited to immortalized cell lines, scientists can produce primary cells. Stem cells, and organoids with appropriate cellular and genetic backgrounds. For instance, human induced pluripotent stem cells (hiPSCs) from people of diverse ethnicities and backgrounds can be used to produce nearly any cell type. From such stem cells, isogenic cell lines can be easily generated, and the genetic variation associated with diseases can be recreated through precise genome engineering. The ability to produce such realistic disease models has greatly enhanced the efficacy of hit validation, enabling potential drugs to be assessed more accurately.
- Drug Discovery

Lead Optimization

- Lead identification/optimization is one of the most important steps in drug development. The chemical structure of the lead compound is used as a starting point for chemical modifications in order to improve potency, selectivity, or pharmacokinetic parameters.

- Once a molecule is identified, the next step is to check its ADMET (Adsorption, Distribution, Metabolism, Excretion and Toxicity) properties. If the molecule has no toxicity and no mutagenicity eight, it has potential for use as a lead molecule. Further optimization gives better quality of lead molecules. These may subsequently be developed as a drug(S).

Drug Discovery and Genome Editing using CRISPR HIT and lead optimization and CRISPR tailored disease Models

CRISPR Tailored Disease Models

- CRISPR-based genome editing is a technology derived from bacteria that is revolutionizing how we study gene function in almost all model organisms, from files to primates.
- CRISPR creates new possibilities for genome manipulation, offering a flexible and user-friendly platform to explode the genetic cause of disease.
- First described as a means to create simple genetic mutants only 3 years ago, CRISPR tools are now being adapted to replace, rearrange, silence, activate, and remodel genomic elements.
- CRISPR changes the way we study biology *in vivo.*

Genomic Targeting of Cas9/sgRNA Complexes

Genomic targeting of Cas9/RNA complexes is mediated by the complementarity between the sgRNA (orange) and a genomic DNA (gDNA) target site (green) that lies immediately upstream of a protospacer adjacent motif (PAM, purple). For Cas9 derived from *Streptococcus pyogenes*, the PAM sequence is NGG, or less commonly, NAG.Cas9 variants from different bacterial species have alternative PAM requirements. Effective recognition and unwinding of the gDNA from the PAM results in double-strand cleavage by two independent nuclease domains in Cas9 (blue arrows).

Applications of CRISPR tools for genome modification

DNA Repair Mechanisms Dictate Genome Editing Outcomes

Following CRISPR/Cas9-induced DNA double-strand breaks, DNA is usually repaired by either homology-directed repair (HDR), or non-homologus end joining (NHEJ). HDR enables the integration of exogenous 'repair templates' such as single-stranded DNA (dsDNA). NHEJ results in the generation of random insertions and deletions (indels) that may disrupt coding regions, or catalyze genomic rearrangements. The cellular preference for HDR or NHEJ following DNA damage can be augmented by small molecules that interfere with each mechanism, and thus bias toward the other. Abbreviations: ATM = Ataxia telangiectasia mutated; ATR = ataxia telangiectasia and Rad3-related; RAD51=Rad51 recombinase defective repair in Chinese hamster cells5; AZT = Azidothymidine; TFT= Trifluridine.

CRISPR Tailored Disease Models

- CRISPR*- Cas9 is the ideal gene-editing system for the establishment of disease-relevant models, as it allows the modification of cells in their pluripotent state.
- However, the single base- pair mutations induced by this system are too small to support the use of screening genes.
- This means that many clones must be sequenced in order to identify those that contain your desired mutation; increasing the time spend optimizing cell culture conditions.
- CRISPR technology can be applied to create new *in vivo* disease models. These tools, derived from an adaptive bacterial immune system, are helping us better model the complexity of human cancer.

CRISPR Tailored Disease Models

- In a comprehensive 2017 review, scientists from the University of California, Berkeley, including co-discoverer of CRISPR-Cas Jennifer Doudna, emphatically concluded that this type of gene editing is "ready to have an immediate impact in real-world drug discovery and development".
- Christof Fellmann, a biotechnologist and co-author of the review, explains that the ability of CRISPR- Cas to help identify target

molecules will have a crucial impact on drug discovery. By using the system to deliberately activate or inhibit genes, researchers can determine the genes and proteins that cause or prevent disease, therefore identifying targets for potential drugs. CRISPR-Cas is also making it easier to create cellular and whole-animal model systems that precisely mimic disease. This is enabling scientists to more accurately verify the safety and efficacy of drugs, which ensures that such models are better predictors of what will happen in clinical trials. As these users are pursued, researchers are also refining and extending the capabilities of CRISPR-Cas to make it an even more powerful editing tool.

CRISPR Tailored Disease Models

- The mechanism that underlies CRISPR-Cas gene editing is relatively simple. A short strand of RNA, tailored to target specific sequences of DNA, is linked to an enzyme that is capable of cutting double-stranded DNA. Cas9, the enzyme to which Moore refers, is to most widely used, but other enzymes are being explored. After the RNA and enzyme are delivered to the cell nucleus, the RNA binds to its complementary DNA sequence, acting as a guide for the enzyme that then chops the DNA. After that crucial cut is made, DNA-repair enzymes in the cell fix the break in a way that either disables or modifies the targeted gene-its activity can be turned up or down, mutations can be introduced, or sections can be inverted.
- The simplicity of using guide RNAs to target any location in the genome is making gene editing accessible to many more researchers. "CRISPR-Cas has taken gene editing out of the hands of those specialists who are expert in complicated molecular biology," says Moore.

CRISPR Tailored Disease Models

- Knock-out screening to identify genes involved in drug resistance is fast becoming one of the exposed most widely used applications of CRISPR-Cas gene editing in drug discovery. Researchers expose a large number of cells to a pool of CRISPR-Cas systems carrying guide RNAs that target various genes. This allows them to generate and select individual cells that each have a specific gene knocked out. The cells are then exposed to chemicals or drugs of interest. Gene that confer resistance to a drug can be identified through cells that become sensitive to such compounds after the CRISPR-Cas treatment. These genes, or the protein they encode, can then be targeted with drugs to get around the problem of resistance.

- Identifying a gene that promotes disease uncovers some obvious targets for drugs development. The simplest candidate drugs bind to and interfere with the proteins encoded by these genes, rather than affect the genes directly. But more-subtle targets for drugs can be revealed by a better understanding of the importance of multiple genes and proteins, their interactions and their mutual regulatory effects. Many diseases, for example, arise when things go wrong in a regulatory pathway that involves a complex network of intracellular interactions. Using CRISPR-Cas to identify, with ease and accuracy, a combination of a gene involved in these networks should offer a more sophisticated approach to treatment.

CRISPR Tailored Disease Models

- The emergence of CRISPR/Cas gene-editing technology with its simplicity, versatility, and efficiency has considerably improved the time frame and the process of creating these modified alleles. The technology has enabled rapid production of knockout, conditional alleles or mice carrying single point mutations, which mimic those in human patients in only several weeks. For the generation of knockout alleles, the microinjection of a single guide RNA (sgRNA) in zygotes is sufficient to create indels in a critical exon, inducing a frameshift mutation and therefore functionally abolishing the gene of interest.
- However, the generation of more complex alleles, including conditionals, is more challenging and requires at least two sgRNA and a repair template in the form of two short single-stranded DNA (ssDNA) repair templates each containing a loxP site. The repair template replaces the endogenous locus of interest by homology-directed repair after Cas cleavage and the double-strand break (DSB) of the region of interest.
- Deskgen

Website: https://www.deskgen.com/landing/

Description

Design any genome editing experiment in any cell line. DESKGEN is organized into two modules: "Experiments" and "Repositories" which together provide genome-editing tools. The software can help to design, build and analyze CRISPR experiments.

- sgRNAcas9

Category:Standalone/command-line

Website:http://www.biootools.com/

Description:

A software package developed for the fast design of CRISPRsgRNA with a minimized off-target effect, in order to improve CRISPR-Cas9 specificity for targeted genome editing and transcriptional control.

- caRpools

Category:Standalone/ command-line

Website: http://github.com/boutroslab/caRpools

Description

This tool provides a complete workflow to analyze CRISPR/Cas9 screens. To further support the analysis of large-scale screens, caRpools integrates screening documentation and the generation of standardized analysis reports. It is user-friendly for both novice and expert users: caRpools' open virtual appliance allows analysis without prior programming knowledge.

- CRISPRseek

Category: Standalone

Website: http://bioconductor.org/packages/release/bioc/html/CRISPRseek.html

Description

The tool is used to identify target-specific gRNAs for CRISPR-Cas9 genome-editing systems. CRISPRseek is a program that simplifies the design of target-specific guide RNA (gRNA) for any CRISPR-Cas9 system with a characterized PAM sequence. It also includes functionalities for score target sites in two related input sequences in order to identify gRNAs.

- CRISPRdirect

Category: Web/online/graphic interface

Website: http://crispr.dbcls.jp/

Description

Consists of a web server for selecting rational CRISPR/Cas targets from an input sequence. CRISPRdirect enables the selection of rational target sequences with minimized off-target sites by performing exhaustive searches against genomic sequences. It is able to investigate the entire genome for perfect matches with each candidate's target sequence. Users can also browse the detailed list of potential off-target sites that have partial complementarity with the selected sequence.

- CRISPRer

Category: Web/online/graphic interface

Website: http://jstacs.de/index.php/CRISPRer

Description:

CRISPRer is a tool for genome-wide selection and assessment of CRISPR/Cas protospacers. It offers three distinct tasks. First, CRISPRer will search for protospacers with unique seed sequences, Second, it will search for seed sequences that are orthogonal to the input genome. Third, it may assess a set of user-provided protospacers by collecting several statistics about genome-wide perfect and imperfect matches to their seed sequences.

- CRISPRTarget

Category:

Web/online/graphic interface

Website: http://bioanalysis.otago.ac.nz/CRISPRTarget

Description

CRISPRTarget tool predicts the most likely targets of CRISPR RNAs. This can be used to discover targets in newly sequenced genomic or metagenomic data. The output provided is either visual in HTML format. The target sequence is displayed as an R-loop, depicting a specified part of the crRNA, as well as the target and non-target strand of the double-stranded target DNA. The target sequence R-loop can be fully reverse complemented when users suspect that the direction of transcription of the CRISPR array starts from the downstream end instead.

- CRISPRfinder

Category:

Web/online/graphic interface

Website: http://crispr.u-psud.fr/Server/

Description

It Investigates clustered regularly interspaced short palindromic repeats (CRISPRs) from genomic sequences. CRISPRFinder recognizes CRISPR arrays using CRISPR repeats. It simplifies the assessment of tentative CRISPR and cas loci. This tool can characterize conserved regions to a single base-pair resolution. It enables the extraction of flanking sequences or spacers and through a desktop version as well as a web application.

- CRISPR gRNA Design tool

Category:

Web/online/graphic interface

Website:https://www.dna20.com/eCommerce/cas9/input

Description

Allows users to design their guide RNA (gRNA). 'CRISPR gRNA Design tool' can target a sequence, a genomic region, or a gene. This tool performs in three steps: (1) selection of the target, (2) results, and (3) delivery options.

Category:

Webserver(Standalone edition available)

Website: http://www.rgenome.net/cas-designer/

Description

It is a user-friendly program to aid researchers in choosing appropriate target sites in a gene of interest for type II CRISPR/Cas-derived RNA-guided endonucleases (RGENs), which are now widely used for biomedical research and biotechnology. Cas-Designer displays the results in an interactive table and provides user-friendly filter functions

11.CRISPR-ERA

Category:

Webserver

Website: http://CRISPR-ERA.stanford.edu

Description

A Fast and Comprehensive Guide RNA Design Tool for Genome Editing, Repression and Activation (CRISPR-ERA).CRISPR-ERA can provide different sgRNA searching approaches for genome editing, such as Cas9 nuclease. In addition, CRISPR-ERA also generates sgRNAs for gene activation or repression using our large-scale database of CRISPRi in different genomes.

12.Cas-OFFinder

Category:

Webserver

Website: http://www.rgenome.net/cas-offinder/

It is an algorithm that searches for potential off-target sites of Cas9 RNA-guided endonucleases. Known to be fast and versatile.

Cas-Designer

13. CRISPRmap

Category:

Webserver

Website:

http://rna.informatik.uni-freiburg.de/CRISPRmap/Input.jsp

Description

CRISPRmap provides detailed insight into repeat conservation and diversity of both bacterial and archaeal systems. It comprises the largest dataset of CRISPRs to date and enables comprehensive independent clustering analyses to determine conserved sequence families, potential structure motifs for endoribonucleases, and evolutionary relationships.

14. CHOPCHOP

Category:

Webserver

Website: https://chopchop.rc.fas.harvard.edu/

CHOPCHOP is a web tool used for selecting the optimum target sites for CRISPR/Cas9- or TALEN-directed mutagenesis. It can be run with few three

basic input options, or with additional advanced parameters.

15.CRISPOR

Category:

Webserver

Website:http://crispor.tefor.net/

CRISPOR assists with guide selection in 120 genomes which include plants and many emerging model organisms, and pre-calculated results for all human coding exons as a UCSC Genome Browser track. Enhanced off-target prediction accuracy.

16. Breaking CAS

Category:

Webserver

Website: http://bioinfogp.cnb.csic.es/tools/breakingcas/index.php

BreakingCAS is a versatile system for detecting putative sgRNA off-targets in CRISPR/Cas applications. Breaking Cas allows working with all eukaryotic genomes available in the ENSEMBL database, for a detailed inspection of the genomic neighborhood of every single potential off-target so as to qualitatively assess them.

CONCLUSION

CRISPR is an emerging new technology and used for genome editing, therapeutic application and applies successfully in clinical implementation. This technology is very effective and safe delivery into the target site. CRISPR-Cas systems classification is introduced to provide an essential template for genome editing and also evolutionary study. However, a new type of CRISPR-Cas system is refined or improved by numerous variants. This technology is still requiring optimization for successful development advancement of this gene editing technology. The continuous advancement of this technology will be involved for improvement in therapeutic applications in many areas. It will allow the discovery of novel drug targets and require considerable research to fully known about the biological mechanisms by CRISPR/Cas9 screens. Moreover, future upgrades of viral and non-viral conveyance strategies will be important to improve in vivo use of CRISPR/Cas9, laying the ground for the helpful utilization of CRISPR later on in respect to therapeutic applications.

REFERENCES

1. Adikusuma, F, Williams, N, Grutzner, F, Hughes, J, Thomas, P. (2017). Targeted Deletion of an Entire Chromosome Using CRISPR/Cas9. Mol. Ther. 25, 1736-1738.
2. Ahmad, G. Amiji, M. (2018). Use of CRISPR/Cas9 gene-editing tools for developing models in drug discovery. Drug Discov. Today, 23, 519-533.
3. Babu M, Beloglazova N, Flick R, Graham C, Skarina T, Nocek B, Gagarinova A, Pogoutse O, Brown G, Binkowski A, Phanse S, Joachimiak A, Koonin EV, Savchenko A, Emili A, Greenblatt J, Edwards AM, Yakunin AF. (2001). A dual function of the CRISPR-Cas system in bacterial antivirus immunity and DNA repair. Mol Microbiol. 79:484-502.
4. Barrangou, R, Fremaux, C, Deveau, H, Richards, M, Boyaval, P, Moineau, S.; Romero, D.A.; Horvath, P. (2007). CRISPR provides acquired resistance against viruses in prokaryotes. Science, 315, 1709-1712.
5. Beloglazova N, Brown G, Zimmerman MD, Proudfoot M, Makarova KS, Kudritska M, Kochinyan S, Wang S, Chruszcz M, Minor W, Koonin EV, Edwards AM, Savchenko A, Yakunin AF (2008). A novel family of sequence-specific endoribonucleases associated with the clustered regularly interspaced short palindromic repeats. J Biol Chem. 283:20361-20371.
6. Bibikova, M.; Carroll, D.; Segal, D.J.; Trautman, J.K.; Smith, J.; Kim, Y.G.; Chandrasegaran, S. (2001). Stimulation of homologous recombination through targeted cleavage by chimeric nucleases. Mol. Cell. Biol. 21, 289-297.
7. Biagioni, A. et al. (2018). Delivery systems of CRISPR/Cas9- based cancer gene therapy. J. Biol. Eng. 12, 33
8. Brendel, J. et al. (2014). A complex of Cas proteins 5, 6, and 7 is required for the biogenesis and stability of clustered regularly interspaced short palindromic repeats (CRISPR)-derived RNAs (crRNAs) in *Haloferax volcanii. J. Biol. Chem.* 289, 7164-7177.
9. Brouns SJ, Jore MM, Lundgren M, Westra ER, Slijkhuis RJ, Snijders AP, Dickman MJ, Makarova KS, Koonin EV, van der Oost J. (2008). Small CRISPR RNAs guide antiviral defense in prokaryotes. Science. 321:960–964.

10. Burstein, D. et al. (2017). New CRISPR-Cas systems from uncultivated microbes. *Nature* 542, 237–241.
11. Carte J, Wang R, Li H, Terns RM, Terns MP (2008). Cas6 is an endoribonuclease that generates guide RNAs for invader defense in prokaryotes. Genes Dev. 22:3489-3496.
12. Casini, A. et al. (2018). A highly specific SpCas9 variant is identified by in vivo screening in yeast. *Nat. Biotechnol.* 36, 265-271.
13. Chen, J. S. et al. (2017). Enhanced proofreading governs CRISPR-Cas9 targeting accuracy. *Nature* 550, 407-410.
14. Cocozaki AI, Ramia NF, Shao Y, Hale CR, Terns RM, Terns MP, Li H (2012). Structure of the Cmr2 subunit of the CRISPR-Cas RNA silencing complex. Structure. 20:545-553.
15. Choi, P.S.; Meyerson, M. (2014). Targeted genomic rearrangements using CRISPR/Cas technology. Nat. Commun. 5, 3728.
16. Chu, V.T. et al. (2015). Increasing the efficiency of homology directed repair for CRISPR-Cas9-induced precise gene editing in mammalian cells. Nat. Biotechnol. 33, 543-548.
17. Cong, L. et al. (2013). Multiplex genome engineering using CRISPR/ Cas systems. Science (80-.) 339, 819-823.
18. Deltcheva E, Chylinski K, Sharma CM, Gonzales K, Chao Y, Pirzada ZA, Eckert MR, Vogel J, Charpentier E. CRISPR RNA maturation by trans-encoded small RNA and host factor RNase III. Nature. 2011; 471:602-607.
19. Deveau, H. et al. (2008). Phage response to CRISPR-encoded resistance in *Streptococcus thermophilus*. *J. Bacteriol.* **190**, 1390-1400.
20. Dominguez, A.A.; Lim, W.A.; Qi, L.S. (2016). Beyond editing: Repurposing CRISPR-Cas9 for precision genome regulation and interrogation. Nat. Rev. Mol. Cell Biol. 17, 5-15.
21. Esvelt, K. M. et al. (2013). Orthogonal Cas9 proteins for RNA-guided gene regulation and editing. *Nat. Methods* **10**, 1116-1121.
22. Fishman-Lobell, J.; Haber, J.E. (1992). Removal of nonhomologous DNA ends in double-strand break recombination: The role of the yeast ultraviolet repair gene RAD1. Science, 258, 480–484.
23. Garneau, J. E. et al. (2010). The CRISPR/Cas bacterial immune system cleaves bacteriophage and plasmid DNA. *Nature* 468, 67-71.

24. Gasiunas, G.; Barrangou, R.; Horvath, P.; Siksnys, V. (2012). Cas9-crRNA ribonucleoprotein complex mediates specific DNA cleavage for adaptive immunity in bacteria. Proc. Natl. Acad. Sci. USA, 109, E2579–E2586.
25. Guichard, S.M. (2017). CRISPR–Cas9 for Drug Discovery in Oncology. In Platform Technologies in Drug Discovery and Validation; Elsevier: Amsterdam, The Netherlands, pp. 61–85.
26. Hale CR, Majumdar S, Elmore J, Pfister N, Compton M, Olson S, Resch AM, Glover CV III, Graveley BR, Terns RM, Terns MP. (2012). Essential features and rational design of CRISPR RNAs that function with the Cas RAMP module complex to cleave RNAs. Mol Cell. 45:292-302.
27. Gao, L. et al. (2017). Engineered Cpf1 variants with altered PAM specificities. *Nat. Biotechnol.* 35, 789-792.
28. Gomaa, A. A. et al. (2014). Programmable removal of bacterial strains by use of genome-targeting CRISPR-Cas systems. *mBio* 5, e00928-00913.
29. Hale CR, Zhao P, Olson S, Duff MO, Graveley BR, Wells L, Terns RM, Terns MP (2009). RNA-guided RNA cleavage by a CRISPR RNA-Cas protein complex. Cell. 139:945-956.
30. Han D, Lehmann K, Krauss G. (2009). SSO1450–a CAS1 protein from *Sulfolobus solfataricus* P2 with high affinity for RNA and DNA. FEBS Lett. 583:1928-1932.
31. Hayes, R. P. et al. (2016). Structural basis for promiscuous PAM recognition in type I-E Cascade from *E. coli*. *Nature* 530, 499-503.
32. Heidenreich, M. and Zhang, F. (2016). Applications of CRISPR-Cas systems in neuroscience. Nat. Rev. Neurosci. 17, 36-44.
33. Haurwitz RE, Jinek M, Wiedenheft B, Zhou K, Doudna JA (2010). Sequence- and structure- specific RNA processing by a CRISPR endonuclease. Science. 329:1355-1358.
34. Hochstrasser, M. L. et al. (2014). CasA mediates Cas3- catalyzed target degradation during CRISPR RNA-guided interference. *Proc. Natl Acad. Sci. USA* 111, 6618–6623.
35. Hou, Z. et al. (2013). Efficient genome engineering in human pluripotent stem cells using Cas9 from *Neisseria meningitidis*. *Proc. Natl Acad. Sci. USA* 110,15644-15649.

36. Hu, J. H. et al. (2018). Evolved Cas9 variants with broad PAM compatibility and high DNA specificity. *Nature* 556, 57-63.
37. Hsu, P. D., Lander, E. S. & Zhang, F. (2014).Development and applications of CRISPR-Cas9 for genome engineering. *Cell* 157, 1262–1278.
38. Jakociunas, T.; Bonde, I.; Herrgård, M.; Harrison, S.J.; Kristensen, M.; Pedersen, L.E.; Jensen, M.K.; Keasling, J.D. (2015). Multiplex metabolic pathway engineering using CRISPR/Cas9 in Saccharomyces cerevisiae. Metab. Eng. 28, 213–222.
39. Jackson, R. N., Lavin, M., Carter, J. & Wiedenheft, B. (2014). Fitting CRISPR-associated Cas3 into the helicase family tree. *Curr. Opin. Struct. Biol.* 24, 106–114.
40. Jinek,M.; Chylinski, K.; Fonfara, I.; Hauer,M.; Doudna, J.A. (2012). Charpentier, E. A programmable dual-RNA-guided DNA endonuclease in adaptive bacterial immunity. Science, 337, 816–821.
41. Jinek, M. et al. (2014). Structures of Cas9 endonucleases reveal RNA-mediated conformational activation. Science 343, 1247997.
42. Jore MM, Lundgren M, van Duijn E, Bultema JB, Westra ER, Waghmare SP, Wiedenheft B, Pul U, Wurm R, Wagner R, Beijer MR, Barendregt A, Zhou K, Snijders AP, Dickman MJ, Doudna JA, Boekema EJ, Heck AJ, van der Oost J, Brouns SJ (2011). Structural basis for CRISPR RNA-guided DNA recognition by Cascade. Nat Struct Mol Biol. 18:529-536.
43. Kim, S. et al. (2014) Highly efficient RNA-guided genome editing in human cells via delivery of purified Cas9 ribonucleoproteins. Genome Res. 24, 1012-1019.
44. Kim, E. et al. (2017). In vivo genome editing with a small Cas9 orthologue derived from *Campylobacter jejuni*. *Nat. Commun.* 8, 14500.
45. Kleinstiver, B. P. et al. (2015). Engineered CRISPR-Cas9 nucleases with altered PAM specificities. *Nature* 523, 481–485.
46. Kleinstiver, B. P. et al. (2016). High-fidelity CRISPR-Cas9 nucleases with no detectable genome-wide off-target effects. *Nature* 529, 490-495.
47. Kleinstiver, B. P. et al. (2019). Engineered CRISPR–Cas12a variants with increased activities and improved targeting ranges for gene, epigenetic and base editing. *Nat. Biotechnol.* 37, 276-282.

48. Kocak, D. D. et al. (2019). Increasing the specificity of CRISPR systems with engineered RNA secondary structures. *Nat. Biotechnol.* https://doi.org/10.1038/ s41587-019-0095-1.

49. Koonin, E. V., Makarova, K. S. & Zhang, F. (2017). Diversity, classification and evolution of CRISPR-Cas systems. *Curr. Opin. Microbiol.* 37, 67-78.

50. Koo Y, Ka D, Kim EJ, Suh N, Bae E (2013). Conservation and variability in the structure and function of the Cas5d endoribonuclease in the CRISPR-mediated microbial immune system. J Mol Biol. 425:3799-3810.

51. Kurata, M.; Yamamoto, K.; Moriarity, B.S.; Kitagawa, M.; Largaespada, D.A. (2018). CRISPR/Cas9 library screening for drug target discovery. J. Hum. Genet. 63, 179-186.

52. Lemos, B.R. et al. (2018). CRISPR/Cas9 cleavages in budding yeast reveal templated insertions and strand-specific insertion/ deletion profiles. Proc. Natl. Acad. Sci. U. S. A. 115, E2040-E2047.

53. Li, W.; Teng, F.; Li, T.; Zhou, Q. (2013). Simultaneous generation and germline transmission of multiple gene mutations in rat using CRISPR-Cas systems. Nat. Biotechnol. 31, 684-686.

54. Liang, F.; Han, M.; Romanienko, P.J.; Jasin, M. (1998). Homology-directed repair is a major double-strand break repair pathway in mammalian cells. Proc. Natl. Acad. Sci. USA 95, 5172-5177.

55. Lin, S. et al. (2014). Enhanced homology-directed human genome engineering by controlled timing of CRISPR/Cas9 delivery. Elife 3, e04766.

56. Liu, J. et al. (2019). Building potent chimeric antigen receptor T cells with CRISPR genome editing. Front. Immunol. 10, 456.

57. Lucas, D. et al. (2017). Utility of CRISPR/Cas9 systems in hematology research. Exp. Hematol. 54, 1-3

58. Makarova, K.S.; Haft, D.H.; Barrangou, R.; Brouns, S.J.J.; Charpentier, E.; Horvath, P.; Moineau, S.; Mojica, F.J.M.; Wolf, Y.I.; Yakunin, A.F.; et al. (2011). Evolution and classification of the CRISPR-Cas systems. Nat. Rev. Microbiol. 9, 467–477.

59. Makarova, K. S. et al. (2015). An updated evolutionary classification of CRISPR-Cas systems. *Nat. Rev. Microbiol.* **13**, 722-736.

60. Makarova KS, Grishin NV, Shabalina SA, Wolf YI, Koonin EV (2006). A putative RNA-interference-based immune system in prokaryotes:

computational analysis of the predicted enzymatic machinery, functional analogies with eukaryotic RNAi, and hypothetical mechanisms of action. Biol Direct. 1:7.

61. Mali, P. et al. (2013). RNA-guided human genome engineering via Cas9. Science (80-.) 339, 823-826.
62. Maruyama, T. et al. (2015). Increasing the efficiency of precise genome editing with CRISPR-Cas9 by inhibition of nonhomologous end joining. Nat. Biotechnol. 33, 538-542.
63. Mojica, F.J.M.; Díez-Villaseñor, C.; García-Martínez, J.; Almendros, C. (2009). Short motif sequences determine the targets of the prokaryotic CRISPR defence system. Microbiology, 155, 733–740.
64. Nam KH, Haitjema C, Liu X, Ding F, Wang H, DeLisa MP, Ke A. (2012). Cas5d protein processes pre-crRNA and assembles into a cascade-like interference complex in subtype I-C/Dvulg CRISPR-Cas system. Structure. 2012; 20:1574-1584.
65. Neggers, J.E.; Vercruysse, T.; Jacquemyn, M.; Vanstreels, E.; Baloglu, E.; Shacham, S.; Crochiere, M.; Landesman, Y.; Daelemans, D. (2015). Identifying drug-target selectivity of small-molecule CRM1/XPO1 inhibitors by CRISPR/Cas9 genome editing. Chem. Biol. 22, 107–116.
66. Niewoehner O, Jinek M, Doudna JA (2014). Evolution of CRISPR RNA recognition and processing by Cas6 endonucleases. Nucleic Acids Res. 42:1341-1353.
67. Nunez JK, Kranzusch PJ, Noeske J, Wright AV, Davies CW, Doudna JA. (2014). Cas1-Cas2 complex formation mediates spacer acquisition during CRISPR-Cas adaptive immunity. Nat Struct Mol Biol. 21:528–534.
68. Ran, F. A. et al. (2015). *In vivo* genome editing using *Staphylococcus aureus* Cas9. *Nature* 520, 186-191.
69. Reeks J, Sokolowski RD, Graham S, Liu H, Naismith JH, White MF (2013). Structure of a dimeric crenarchaeal Cas6 enzyme with an atypical active site for CRISPR RNA processing. Biochem J. 452:223-230.
70. Richter C, Gristwood T, Clulow JS, Fineran PC. (2012). In vivo protein interactions and complex formation in the Pectobacterium atro-septicum subtype I-F CRISPR/Cas System. PLoS One. 7:e49549.

71. Richter H, Lange SJ, Backofen R, Randau L. (2013). Comparative analysis of Cas6b processing and CRISPR RNA stability. RNA Biol. 10:700-707.

72. Richardson, C.D. et al. (2016). Enhancing homology-directed genome editing by catalytically active and inactive CRISPRCas9 using asymmetric donor DNA. Nat. Biotechnol. 34, 339-344.

73. Rouillon C, Zhou M, Zhang J, Politis A, Beilsten-Edmands V, Cannone G, Graham S, Robinson CV, Spagnolo L, White MF (2013). Structure of the CRISPR interference complex CSM reveals key similarities with cascade. Mol Cell. 52:124-134.

74. Sashital DG, Wiedenheft B, Doudna JA. (2012). Mechanism of foreign DNA selection in a bacterial adaptive immune system. Mol Cell. 46:606-615.

75. Scott, A. (2018). How CRISPR is transforming drug discovery. Nature, 555, S10–S11.

76. Shmakov, S. et al. (2017). Diversity and evolution of class 2 CRISPR-Cas systems. *Nat. Rev. Microbiol.* **15**, 169-182.

77. Sinkunas, T. et al. (2011). Cas3 is a single-stranded DNA nuclease and ATP-dependent helicase in the CRISPR/ Cas immune system. *EMBO J.* 30, 1335-1342.

78. Slaymaker, I. M. et al. (2016). Rationally engineered Cas9 nucleases with improved specificity. *Science* **351**, 84-88.

79. Smurnyy, Y.; Cai, M.; Wu, H.; McWhinnie, E.; Tallarico, J.A.; Yang, Y.; Feng, Y. (20014). DNA sequencing and CRISPR-Cas9 gene editing for target validation in mammalian cells. Nat. Chem. Biol. 10, 623-625.

80. Spilman M, Cocozaki A, Hale C, Shao Y, Ramia N, Terns R, Terns M, Li H, Stagg S. (2013). Structure of an RNA silencing complex of the CRISPR-Cas immune system. Mol Cell. 52:146–152.

81. Staals RH, Agari Y, Maki-Yonekura S, Zhu Y, Taylor DW, van Duijn E, Barendregt A, Vlot M, Koehorst JJ, Sakamoto K, Masuda A, Dohmae N, Schaap PJ, Doudna JA, Heck AJ, Yonekura K, van der Oost J, Shinkai A. (2013). Structure and activity of the RNA-targeting Type III-B CRISPR-Cas complex of *Thermus thermophilus*. Mol Cell. 52:135–145.

82. Sternberg, S.H. et al. (2014). DNA interrogation by the CRISPR RNA-guided endonuclease Cas9. Nature 507, 62-67

83. Torres, R.; Martin, M.C.; Garcia, A.; Cigudosa, J.C.; Ramirez, J.C. (2014). Rodriguez-Perales, S. Engineering human tumour-associated chromosomal translocations with the RNA-guided CRISPR-Cas9 system. Nat. Commun. 5, 3964.
84. Tsai, S.Q.; Joung, J.K. (2016). Defining and improving the genome-wide specificities of CRISPR-Cas9 nucleases. Nat. Rev. Genet. 17, 300-312.
85. Van Duijn E, Barbu IM, Barendregt A, Jore MM, Wiedenheft B, Lundgren M, Westra ER, Brouns SJ, Doudna JA, van der Oost J, Heck AJ. (2012). Native tandem and ion mobility mass spectrometry highlight structural and modular similarities in clustered-regularly-interspaced shot-palindromic-repeats (CRISPR)-associated protein complexes from *Escherichia coli* and *Pseudomonas aeruginosa*. Mol Cell Proteomics. 11:1430-1441.
86. Vakulskas, C. A. et al. (2018). A high-fidelity Cas9 mutant delivered as a ribonucleoprotein complex enables efficient gene editing in human hematopoietic stem and progenitor cells. *Nat. Med.* 24, 1216-1224.
87. Wang, H.; Yang, H.; Shivalila, C.S.; Dawlaty, M.M.; Cheng, A.W.; Zhang, F.; Jaenisch, R. (2013). One-step generation of mice carrying mutations in multiple genes by CRISPR/Cas-mediated genome engineering. Cell 153, 910-918.
88. Wiedenheft B, Zhou K, Jinek M, Coyle SM, Ma W, Doudna JA (2009). Structural basis for DNase activity of a conserved protein implicated in CRISPR-mediated genome defense. Structure. 17:904-912.
89. Wiedenheft B, van Duijn E, Bultema JB, Waghmare SP, Zhou K, Barendregt A, Westphal W, Heck AJ, Boekema EJ, Dickman MJ, Doudna JA. (2011). RNA-guided complex from a bacterial immune system enhances target recognition through seed sequence interactions. Proc Natl Acad Sci U S A. 108:10092-10097.
90. Westra, E. R. et al. (2012). CRISPR immunity relies on the consecutive binding and degradation of negatively supercoiled invader DNA by Cascade and Cas3. *Mol. Cell* 46, 595–605.
91. Yamada, M. et al. (2017). Crystal structure of the minimal Cas9 from *Campylobacter jejuni* reveals the molecular diversity in the CRISPR-Cas9 systems. *Mol. Cell* 65, 1109-1121.
92. Yosef I, Goren MG, Qimron U. (2012). Proteins and DNA elements essential for the CRISPR adaptation process in Escherichia coli. Nucleic Acids Res. 40:5569–5576.

93. Yin, H. et al. (2019). CRISPR-Cas: a tool for cancer research and therapeutics. Nat. Rev. Clin. Oncol. 16, 281-295.

94. Zhang J, Kasciukovic T, White MF (2012). The CRISPR associated protein Cas4 Is a 5′ to 3′ DNA exonuclease with an iron-sulfur cluster. PLoS One. 7:e47232.

95. Zetsche, B. et al. (2015). Cpf1 is a single RNA-guided endonuclease of a class 2 CRISPR-Cas system. *Cell* **163**, 759-771.

96. Zetsche, B. et al. (2017). Multiplex gene editing by CRISPR-Cpf1 using a single crRNA array. *Nat. Biotechnol.* **35**, 31-34.

97. Zhang J, Rouillon C, Kerou M, Reeks J, Brugger K, Graham S, Reimann J, Cannone G, Liu H, Albers SV, Naismith JH, Spagnolo L, White MF. (2012). Structure and mechanism of the CMR complex for CRISPR-mediated antiviral immunity. Mol Cell. 45:303–313.

98. Zhang Y, Heidrich N, Ampattu BJ, Gunderson CW, Seifert HS, Schoen C, Vogel J, Sontheimer EJ. (2013). Processing-independent CRISPR RNAs limit natural transformation in *Neisseria meningitidis*. Mol Cell. 50:488-503.

99. Zuo, Z. and Liu, J. (2016). Cas9-catalyzed DNA cleavage generates staggered ends: evidence from molecular dynamics simulations. Sci. Rep. 6, 37584.

PART-III

LABORATORY REQUIREMENTS

CHAPTER 6

VIROLOGY LABORATORY REQUIREMENTS AND SAFETY RULES & REGULATIONS

INTRODUCTION

Microbiology is the study of microbes (microorganisms), organisms so small that a microscope is needed to study them, in other words, the study of organisms in the microscopic range. Microbes include bacteria, archaea, algae, yeast, fungi, viroids, prions and protozoa. Thus whereas studying microbiology we generally are working with unseen living forms of life. Since the techniques employed in microbiology may be new to many of you, you have to develop perfection. The expertise in these techniques will determine not only the success or failure of your scientific probing but these will be essential to protect you and others around you against potential hazards. You may and may not be working with pathogenic microorganisms but it is essential to observe certain protective measures while working in a microbiology laboratory.

It has been estimated that thousands of infections have been acquired in the microbiology laboratory and many people have died because of such infections. The most common laboratory-acquired microbial infections are: typhoid fever (*Salmonella typhi*) and brucellosis (*Brucella melitensis*) (bacteria), histoplasmosis (*Histoplasma capsulatum*) (molds), hepatitis B virus (viruses). It has been said that all microorganisms should be treated as potential pathogens i.e organisms capable of producing disease. Thus personnel safety should be a major concern in all laboratories. The materials such as live cultures, inoculated culture media, used hypodermic needles, and patient specimens require specific precautions for safe handling.

BASIC REQUIREMENTS IN A MICROBIOLOGY LABORATORY

It is a specialized branch of biological sciences and microbes play a vital role in the ecology of life on earth. According to the scientist Louis Pasteur, Life would not long remain possible in the absence of microbes. However, some microorganisms cause disease in humans, other animals and plants. There is an all sciences need to know about equipment, tools, glassware, and miscellaneous items. There is much of which can be found in any biological laboratory. Here we are listed of basic requirements that a microbiologist requires in his laboratory for microscopic examination, isolation or culturing and identification of microorganisms as well as to learn its structure, function and application.

Equipment and Instruments

- Oven
- Water Bath
- Laminar Air Flow
- Bunsen Burner
- Spirit Lamp
- Microscope
- Immersion Oil
- Constant Water bath
- Incubators
- Refrigerator
- Autoclave
- Homogenizers
- Balances
- Camera Lucida
- Photomicrographic Camera
- Autoclave

- Quebec Colony Counter
- Spectrophotometer
- Scanning Electron Microscope
- Centrifuge
- pH Meter
- Hot Plate
- Heater

Tools

- Thermometers
- Transfer Needle
- Burette Setup
- Inoculating Loop
- Stage Micrometer
- Ocular Microscope
- Hemostat
- Dissecting Needles
- Forceps
- Scissors
- Hemostat

Glassware

- Petri plates
- Glass Cover Slips
- Concave Slides
- Conical Flasks
- Beakers
- Funnels
- Graduated Cylinders
- Pipettes
- Culture Tubes Without Screw Caps and With Caps
- Dropper Bottles

- Screw-Capped Bottles for Stock Reagents
- Glass Microscope Slides

Miscellaneous

- Cotton Gauze Masks
- Pipette
- Petri Dish
- Muslin Cloth
- Culture Media
- Test Tube Rack
- Cotton Plugs
- Wooden Sticks
- Marking Pen
- Stains
- Distilled Water
- Syringes and Needles
- Butter Paper
- Rubber Bands
- Blotting Papers
- Lens Tissue
- Gloves
- Nail Polish/ Wax for Sealing
- Tape For Sealing
- Discard Containers

Safety in a microbiological laboratory significantly differs from so as to in other laboratories because 'addition to hazardous chemicals, substances and operations that poses a laboratory work-related to risk' there is a risk of infection when working with microbes. The presence and working with infectious materials in a microbiological laboratory, i.e. the possibility of acquiring laboratory-associated infection, assumes that control methods. These control measures first take into account the knowingly or unknowingly handled infectious agents. The approval of laboratory practices is the safety equipment used in good laboratory practices. The containment moves toward are the level by which laboratory workers are aware of the risk of infection.

Hence, safety programs and safety management are organized with these questions in mind.

The most common behavior of experience to infectious agents are percutaneous inoculation (by injuries caused by sharp contaminated and animal bites and scratches), aerosol inhalation (as a result of spills, or caused by sprays related with work events, e.g. vortexing; the simple opening of a Petri dish culture of a sporulating fungus, work with lyophilized cultures), and ingestion (during mouth pipetting, or by eating or drinking in the laboratory). Since the infectious dose of a disease-causing agent is critical, the upper concentration of microorganisms associated with certain research procedures (e.g. cultivation) increases the risk of infections. The infectious dose is the amount of microbial cells that cause an acute infection in humans always such as "*Vibrio cholera*" strains cause disease when ingesting only (10 cells), while some "*Escherichia coli*" strains ($>10^6$ cells) are needed for disease stimulation.

In the laboratory, researchers and students are exposed to the maximum risk. However, one has to take an explanation about the exposure of the cleaning and protection to staff. The above-mentioned laboratory employees are frequently understood to be healthy individuals in risk assessments. On the other hand, there are health status conditions, which increase the risk of infections. Different life stages, some diseases and the use of certain medications influence the host's defense such as e.g. pregnancy with the threat of fetal or congenital infection, allergic hypersensitivity, immunodeficiency caused by e.g. diabetes mellitus, cancer chemotherapy. Furthermore, working in a laboratory can result in allergic reactions.

SAFETY GUIDELINES

The rules and regulations of any science laboratory have been developed for the safety and protection of all the persons which are present there. The teacher provides extra safety rules for particular experiments. While conducting experiments in the lab these extra rules must be followed seriously by everyone and every time. All instructions should be noted properly, understood correctly and to be pursued sincerely all the time in the lab. Here are some basic guidelines discussed below which we must follow:

- Only those experiments should be performed which are instructed by your instructor.

- Follow all the instructions carefully, both verbal and written. Those experiments which are not instructed by the teacher are not allowed in the laboratory.
- Wash your hands before starting and after finishing any experiment with detergent (dettol).
- Carry a permanent marker pen and a small napkin with yourself.
- After entering a science laboratory, avoid touching any chemicals, equipment, or any other things until or unless you are instructed to do so.
- Eating or drinking or chewing anything in the laboratory is strictly prohibited.
- The use of laboratory glassware as a container for food or beverages may cause serious infection or poisoning.
- Never try to work on your own in the science laboratory.
- Understand your work clearly before starting an experiment in the laboratory. If there is any difficulty to understand any part of an experiment, ask the teacher before starting that experiment.
- Don't make fun in the laboratory.
- Practical jokes, horseplay, and pranks are proved to be very dangerous.
- The working area should be well-ventilated and clean.
- Report the instructor immediately if we saw any unsafe conditions.
- While using chemicals or any reagent keep your dirty hands away from eyes, face, mouth or any other body part.
- Closed water taps properly.
- Make your washing area neat and clean after completing your experiments.
- Personal monitoring of experiments is necessary at all times. Do not roam around in the laboratory, divert or startle other students or interfere with their experiments.
- All the operating procedures and locations should be known, like safety equipment which includes: fire extinguisher and first aid kit.
- We should know where the exits and fire alarm are located.
- We should know clearly what to do if there is any fire drill during

experiment time; all the containers of chemicals must be closed, and all the electrical equipment should turn off.

Safe Clothing

Proper dressing during experiments in the laboratory is necessary. There are three major things that are counted in safety clothing, such as lab coats, safety goggles, and hand gloves. Open hairs, dangling jewelry (chains or earrings), and loose clothing are definitely serious hazards in the science laboratory. Long hairs must be tightly tied back, hanging jewelry and loose clothing must be avoided. Shoes must be removed outside the laboratory.

- A neat and clean, properly autoclaved laboratory coat is compulsory in laboratory experiments (Figure 6.1).
- One should wear safety goggles if he or she performed experiments related to ultraviolet radiation or activity emitting fume.
- One must avoid the use of contact lenses in the laboratory.

Figure 6.1. Wearing lab coats, safety goggles, and hand gloves

Accidents and Fire

- Report immediately any accidents like cut, burn and spill to the teacher or laboratory instructor.
- Either accident occurred is small or big one should report compulsorily to his/her teacher or laboratory instructor without any panic or hesitation. (Figure 6.2).
- For heat burns, chill the affected part with ice immediately.

- Never leave your cut surfaces open, it must be covered with the bandage.
- If you or your laboratory partner gets hurt by any means, immediately (and loudly) call your teacher.
- If any chemical is a splash on your eye or skin wash immediately with cold running water several times.
- If the problem still persists, visit a nearby doctor.
- For larger spills on the body, use a shower immediately.
- If gas burns from a leak, turn off the gas tap.
- If you smell a leakage of gas, immediately open the doors and windows, check all gas taps and tighten. The regulator/burner switches off the gas.
- Don't switch on any electrical types of equipment or burn a matchstick until the room is free of gas.
- In case of minor fire occurrence use fire extinguishers in the laboratory to extinguish it.
- In case of a large fire occurring in the laboratory itself, leave the room quickly after turning all gas burners off and unplugging the electrical equipment.
- Dial fire helpline number to call the fire service department.

Burns

The fire is stopped with a blanket. The areas burned should be enclosed together with this moist, cool compression or held under running cold water waiting for the ambulance arrives.

Figure 6.2. Laboratory accidents

Chemical Handling

All the chemicals in the laboratory are dangerous (Figure 6.3). Avoid handling chemicals with bare hands. Always use hand gloves. While making an observation, maintain a reasonable safe distance from the specimen and do not smell or try to taste any chemicals.

- Check carefully all the labels of chemical bottles before using them.
- This must be a precautionary step to be taken care of which should be applied before every experiment or activity.
- Take out only as much chemical as you required.
- Do not put back unused chemicals in the container to avoid spoilage of all chemicals in the container.
- Materials susceptible to spontaneous combustion due to contact with a nitric acid include oily rags, dust accumulations, and many organic materials. Routes of entry to the body include inhalation and ingestion. Skin contact results in severe irritation and burns.
- Ethidium bromide (Et-Br) is a carcinogenic compound, which is widely used in molecular analysis. It is mandatory to wear gloves while using this chemical.
- Acids require careful handling and mixing of acid and water also requires due attention. Sulfuric acid causes serious sores.
- Always add acid to water, instead of adding water to acid for mixing. In details (First pour water in the beaker and thereafter pour acid over water and maintain the volume as per requirements of the experiment.
- Mercuric chloride ($HgCl_2$) is a lethal chemical that is commonly used in tissue culture. If it accidentally gets ingested or inhaled by any person, it causes immediate death.
- So one should take utmost safety while handling.
- Liquid N_2 (Nitrogen) at -196°C temperature, may cause serious ice burn on the skin.

Figure 6.3. Laboratory chemicals (dangerous)

Cleaning of Glassware and Handling

Never use broken glassware (Figure 6.4). If any glassware gets broken during an experiment, do not pick it up with your bare hands instead use a dustpan and use a brush to clean them.

- Glasswares should be examined carefully before use.
- Place new glassware like test tubes, Petri dishes, pipettes and flasks into 1% solution of a glass cleaner such as teepol/liquid detergent and heat to boil.
- Remove the glassware and rinse thoroughly with warm tap water.
- Allow the glassware to dry before putting it to use.
- All the glassware must be sterilized in an autoclave prior to washing.
- Do not use immediate cold water on any hot glassware, the glassware may break.
- Always use a mouth pipette with high precaution.
- Have a preference to use micropipettes or use a rubber bulb to suck the chemical.
- Never wipe plastic wares with alcohol, it causes cracking of plastic.

- Before using any instrument, read its handling, principle and precautions carefully.
- We should always be aware of the highest and the lowest limit of the instrument. For eg about a centrifuge (we must be aware of its higher capacity of rotation), weighing machine (highest and lowest limit of weighing) , etc.
- Immerse new glass slides in ethanol containing hydrochloride acid (3%) for several hours to get it properly sterilized.
- Remove the slides and rinse thoroughly with tap water and finally with distilled water.
- Wipe the slides with a soft cotton cloth to make them dry.
- Keep the slides in a covered jar till the experiments are over.
- Place the used pipettes in 3% Lysol solution.
- If necessary keep them immersed in chromic acid overnight, wash them first with tap water and then with distilled or deionized water.

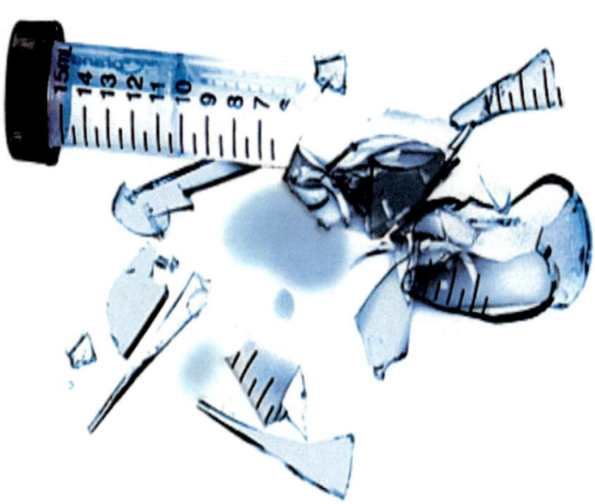

Figure 6.4. Laboratory broken glassware

Waste Disposal

Each and every student, researcher or lab technician should necessarily know to dispose of waste chemicals, used syringes, heavy metals, nanoparticles, lethal compounds ,etc (Figure 6.5). Every chemical causes some reaction with water, air or with some other chemicals or substances, so it is mandatory to know the nature of the waste before disposing of it openly.

- Petri dishes must be placed in a plastic bag to be autoclaved.
- Do not try to pick up the broken glass fragments with your bare fingers, instead use gloves.
- It should be swept up into a dustpan and always be placed in the container reserved for broken glass.
- Used pipettes must be placed in a wire basket to be autoclaved.
- Never place the contaminated material in a wastebasket.

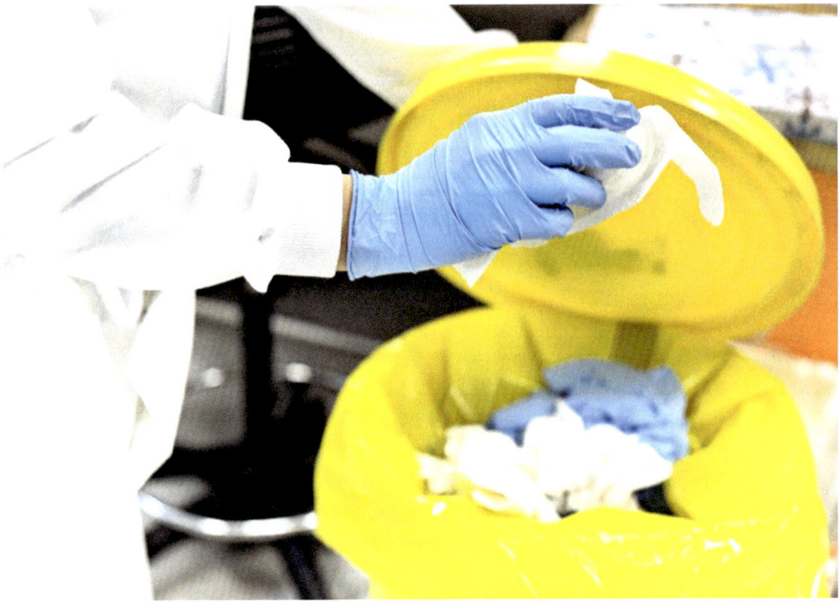

Figure 6.5. Disposal of Waste gloves

Recording Practical Work

All the practical work should be recorded thoroughly in the workbook under the following heads (Figure 6.6)

- Name of the paper
- Name of the teacher
- Date and title of the experiment
- Aim or objective of the experiment

- Materials required
- Details of procedure/methodology
- Observation made
- Diagrams/drawing/comments on results
- Difficulties faced and tips for successful experiments
- Suggestions for further experimentation

Figure 6.6. Figure showing the recording of an experiment in the workbook

Emergency and First Aid Guide

The laboratory should have a first aid kit and stored it into the close of the sink. Optional contents add in such as (sterile bandage adsorbent gauze, crepe bandage, adhesive bandage, compression bandage, scissors, medical tape, disposable gloves, resuscitation pack, betadine, activated carbon tablets, aqueous hydrochloric acid solution (10%), 2 % boric acid solution in dropper dispenser, 2 % borax solution in dropper dispenser, 3 % sodium hydrogen carbonate solution, 3 % boric acid solution, meat extract powder ,etc).

In case of emergency, remain calm. Alert classmates to evacuate the area informs the instructor/local rescue personnel.

Call for emergency or ambulance (call 104 or 112). When talking to the emergency/ambulance personnel follow:

- First, give your name and the telephone number from which you are calling.
- Indicate the exact location and give help on how to reach it on the campus within the building.
- Describe briefly but correctly the number of victims, the emergency conditions of the victim (conscious or not, bleeding, burned, pains, etc.).
- Give indication if there is a need for the fire brigade (in this case you do not need to call them separately, the ambulance personnel will alert them).
- Do not be suspended up, you can ask for first-aid instructions over the telephone, furthermore, additional information may be needed.

Bleeding

Always wear gloves when treating a bleeding person or getting in contact with objects contaminated with blood. Gloves should not be reusable. In the case of "external bleeding" apply constant firm pressure on the wound using e.g. sterile adsorbent bandage, or even with your gloved hands. Concurrently, turn the victim in a position where the place of bleeding can be raised above his/her heart. Internal bleeding can be a similarly serious problem. Coughing or vomiting blood is sign of internal bleeding. In the case of extensive bleeding, lay the victim down and raise his/her legs to "approx. 30 cm". Do not give food or drink.

Unconscious Condition

Unconsciousness may be caused by different diseases and injuries. It refers to a state when respiration and circulation function, but it is impossible to contact the victim. First, try to establish contact by shouting at him/her by shaking the shoulder. Calm down the status of the person's "respiration and circulation" by mouth and pharyngeal toilette, turn the victim on the side and cover him/her in order to keep him/her warm. Open the airways by carefully lifting the neck of the victim.

Heart Attack

The signs and symptoms of a heart attack include

- pressing chest pain

- impeded respiration
- full sweating
- nausea or vomiting
- dizziness
- feeling faint
- weakness
- anxiety

If any of these symptoms occur, call the ambulance immediately. In the meantime, loosen clothing and support the person that help is on its way. Position the victim into a half-sitting position to help to breathe.

BIOSAFETY IN MICROBIOLOGY LAB (BSML)

Each laboratory should develop or adopt a biosafety or operations manual, which identifies the hazards that will or may be encountered, and which specifies the practices and procedures designed to minimize or eliminate risks. The personally advised about special hazards and should be required to read and follow the required practices and procedures. A scientist trained and knowledgeable in appropriate laboratory techniques, safety procedures and hazards associated with handling infectious agents must supervise laboratory activities. Microbiology lab is special often unique, work environments that may pose identifiably infectious disease risks to the person in or near them. Therefore, safety in the microbiology lab is very important.

Safety procedures and precautions to be followed in the microbiology laboratory should be designed to:

- Restrict microorganisms present in specimens or cultures to the vessels in which they are contained
- Prevent environmental microorganisms (normally present on hand, hair, clothing, laboratory benches or in the air) from entering specimens or cultures and interfering with the results of the studies.

The four biological safety level (BSL) categories are mainly based on the harshness of the disease they cause and their transmission route and since airborne transmission (Figure 6.7) is the most complicated to control. Laboratory facilities are classified into four safety levels according to the agent.

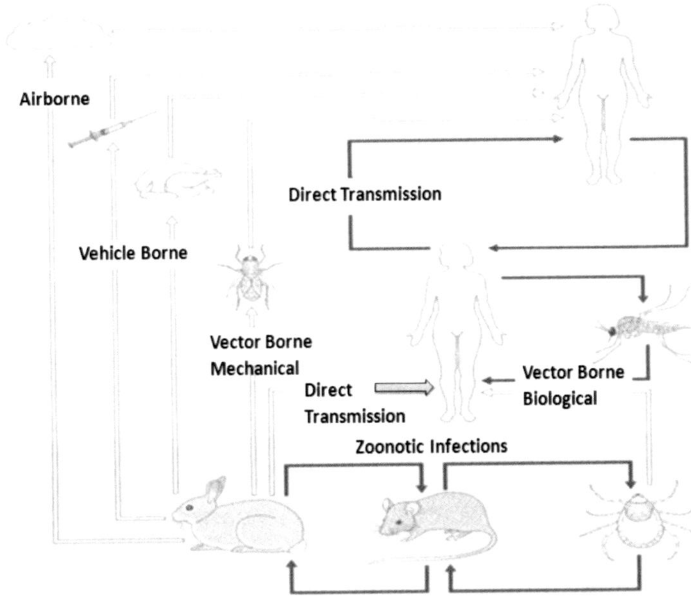

Figure 6.7. Route transmission of disease in human

Organisms in BSL 1 are not known to cause any disease in healthy adults. BSL 1 services are sufficient for teaching laboratories at post-secondary or undergraduate training levels. In such laboratories, only a sink to wash hands for decontamination is needed. Microbes in BSL 2 are transmitted with ingestion, or via contact with mucous membranes. However, their high concentration may result in the transmission of high infectious. Individual protecting equipment should be used (laboratory coats, glasses and goggles, gloves). Microbes ranked as BSL 3 cause disease in humans and explicitly spread airborne (with low infectious dose). Microbes or samples that are verified or only supposed to have a high risk of causing serious or even fatal disease in humans, independently of the transmission route, are categorized as BSL4.

Biosafety Levels (BSL)

The concept of the bio-safety level was first defined by the Center for Disease Control and Prevention (CDCP), United States. The bio-safety level is defined as a set of precautions required to save researchers from bio-contaminants by using some laboratory facilities. Contaminants are divided into some levels according to their effects, so accordingly the set of precautions are prepared. There are 4 different bio-safety levels as kept in sequence from the lowest biosafety level 1 (BSL-1) to the highest at level 4 (BSL-4) (Figure 6.8). Four biosafety levels are described, which consist of combinations of laboratory practices and techniques, safety equipment and laboratory facilities. The biosafety levels are accompanied by precautions to be observed in the microbiology laboratory.

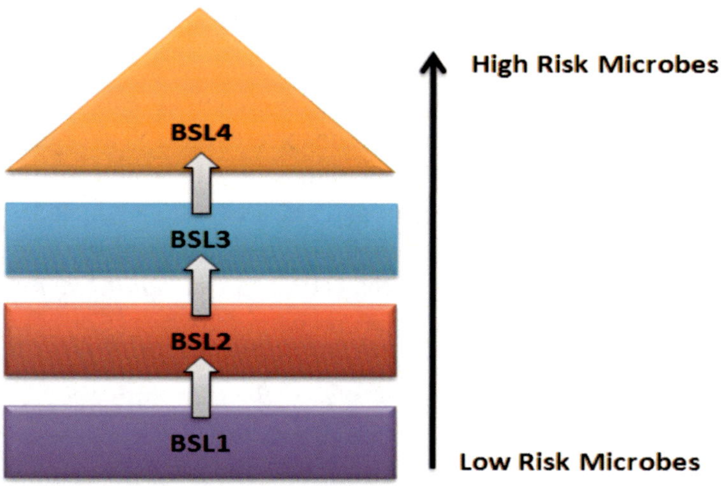

Figure 6.8. Biosafety levels

Biosafety Level 1 (BSL-1)

This is the first biosafety level for practices, safety equipment and facilities are appropriate for undergraduate and secondary educational training and teaching laboratories and other facilities in which work is done with defined and characterized strains of viable microorganisms not known to cause disease in healthy adult humans. BSL-1 is the basic bio-safety level, which has minimal precautions. It deals with those bioagents which generally do not cause any harm to human beings. In other words, it includes those agents which should pose a minimal potential hazard to the laboratory environment and the human being present there. In this level, precautions to be followed less in comparison to other levels. In BSL-1, there should be a door in the laboratories which can be closed or locked to limit access. However, it is not necessary to isolate BSL-1from the general institute building. BSL-1 deals with the different kinds of non-pathogenic microorganisms such as *Escherichia coli*, *Saccharomyces cerevisiae*, *Bacillus subtilis* and some other organisms not suspected to cause any disease to humans, or some others microbes *Adeno-associated viruses 1-4 T4 Bacteriophage, Rhizopus stolonifer, Candida albicans and Pseudomonas* (GEM).

Biosafety Level 2 (BSL-2)

In BSL-2 safety precautions adopted are higher in comparison to BSL-1 such as researchers are required to have specific training to handle pathogenic agents, hence these researchers are trained by qualified which are directed by trained scientists. Access to BSL-2 is limited when particular research is going on. Certain procedures which include infectious agents or any other physical containment equipment are conducted in biological safety cabinets such as a laminar hood. BSL-2 involves those agents which are a moderate potential hazard to the environment and researchers. This includes several microbes such as Hepatitis A, B, and C viruses, pathogenic *Escherichia coli*, human immunodeficiency virus (HIV), *Salmonella*, *Staphylococcus aureus*, *Toxoplasma gondii and Plasmodium falciparum, Cornybacterium, diphtheria, Neisseria gonorrhea, Vibrio cholera, Klebsiella sp, Proteus, Hepatitis A,B,C, Cryptococcus neoformans, Bacillus Anthracis, Bordetella pertusis* ,etc.

Biosafety Level 3 (BSL-3)

Higher safety precautions are adopted in BSL-3 due to its involvement with serious and potentially lethal microbes which can cause disease via the inhalation route. Laboratory type in BSL-3 is research, diagnostic, clinical or production facilities. All the above precautions which are followed in BSL-1 and BSL-2 are followed with some more specific safety measures such as all researchers are provided relevant immunizations and medical surveillance to reduce the risk of any unnoticed accidents. All procedures which involve infectious agents must be done within a laminar hood. Researchers must wear protective clothing such as safety gowns with caps and solid gloves. This clothing are restricted to be worn outside of the laboratory and must be discarded (or decontaminated) after each procedure. The entrance of BSL-3 must be separated from other general areas of the building to restrict the flow of traffic. As well, the doors must be in the form of two sets of self-closing doors (to reduce the escaping of contaminant aerosols). In BSL-3 the laboratory area should be less cozy or less furniture, which may help in easy cleaning. The Carpets and any other seams in the floors, walls, and ceilings are prohibited in this area. Additionally, laboratory windows must be sealed, and a proper ventilation system is installed which forces air to flow out. Before re-circulation of air from the laboratory, it must be filtered. BSL-3 is commonly in involved research and diagnostic work which deals with those microbes that can be transmitted by aerosols and may cause severe diseases such as *Chlamydia psittaci, Mycobacterium tuberculosis, Several Species of Brucella, West Nile Virus, Chikungunya, Yellow Fever Virus, SARS Coronavirus, Francisella Tularensis, Eastern Equine Encephalitis Virus, Coxiella burnetii, Rift Valley Fever Virus, Rickettsia rickettsii, Venezuelan Equine Encephalitis Virus, Hanta virus, SLE virus, VEE virus, HIV Coccidiodes, Yimmitis, Prions, Brucella abortus, Mycobacterium bovis and Pasteurella multocida.*

Biosafety Level 4 (BSL-4)

BSL-4 has the highest level of biosafety precautions. It deals with the use of lethal disease-causing agents for which there are no proper treatments or vaccinations are available. These agents can easily be aerosol transmitted within the laboratory area and cause severe to fatal disease in humans. BSL-4 follows all the safety precautions of BSL-1 to BSL-3 with all other higher level of precautions. BSL-3 laboratories are generally separated from the general building and are in the form of a cabinet or protective suit laboratories. In the cabinet form of laboratories, all the work must be done within a BSL-3 safety cabinet. Materials that are leaving the cabinet must be decontaminated by passing through an autoclave. The cabinets are prepared like that these can be easily cleaned.

Additionally, all the furniture and slabs in the cabinet must be free from sharp edges, which help in reducing the risk of damage to the gloves and safety clothes. After completion of the procedure, researchers must pass through a chemical shower for decontamination before exit from the BSL-4 laboratory. Additionally, there is a separate room for removing the pressure suit, followed by a personal shower. Entry in the BSL-4 laboratory is restricted to trained and authorized individuals. All thc pcrsons who either enter or exit from the laboratory must be under safety surveillance. BSL-4 laboratories are mainly used for diagnostic and research work. It deals with easily transmitted pathogens which can cause fatal disease. These include the *Lassa virus, Ebola virus, Marburg virus and Crimean-Congo hemorrhagic fever virus. Some other pathogens are Nipah virus, Hendra virus, and Flaviviruses, Herpesvirus simiae and Haemorrhagic viruses.*

ETHICS

Ethics are standards of conduct that prescribe behavior. Ethics is a branch of philosophy that involves the concepts of what is right and what is wrong. Before describing ethical issues of laboratory here are some important terms which we should know. Researchers should not misrepresent falsify or fabricate the original data or results. Research should be pure, and the researcher should be unbiased, truthful and objective in the complete process of research. A researcher should not prejudice. Scientists or researchers should always be careful throughout the experiment and data interpretation. They should avoid errors during experiments and especially in presenting their results. They should also avoid conflicts of interest, self-deception, and biases. Research should be open. The researcher should share their ideas, methods, tools, techniques, data and results. They should allow other researchers to review their work and gave their opinions and criticize openly. Scientists should be free to think in every aspect of science and made their hypothesis on any problem.

The research should always be open to new ideas and criticize the old ones. Research credit should be given to that person or source where credit is due but where it is not due there is no need to give credit. It means we should always give credit to a person or resource or institute which helps in our research at any point. Credit is not based on the conflict of interest or any biases. Science is not only to invent something new and do something different. It always comes with the education of next-generation scientists, how to conduct good science. Their experience and work should educate and motivates the public about science. Science must be always used for the social benefit. Scientists should avoid causing harms to society by their invention or research. Scientists should inform the public about their research and their future consequences because they are responsible for these consequences. If any scientist breaks the laws, he or she can be arrested and their funding sources will be ceased. In science, there should no biases or discrimination regarding opportunities. Researchers should treat their colleagues with respect. Respect in the sense of do not harm each other in any manner either physically or psychologically. Do not cause any hindrance in other's research or destroy their data or results.

VIVA QUESTIONS

1. Name a few laboratory-acquired microbial infections?
2. What is the purpose of wearing an apron in a microbiology laboratory?
3. Name a few disinfections used in a microbiology lab?
4. What disinfections are used for spilled culture?
5. What precautions are taken to avoid laboratory-acquired infections?
6. How the cultures are to be disposed of?
7. What is ethics in a science laboratory?

CHAPTER 7

pH METER

INTRODUCTION

A pH meter is an instrument used to measure the pH (acidity and alkalinity) of a solution, of various foods, juices as well as for setting pH of culture media and biochemical activities of microorganisms (Figure 7.1). It is expressed as a number from 0 to 14 categorized into three types: neutral (pH 7), acidic (pH < 7) and alkaline (pH > 7).

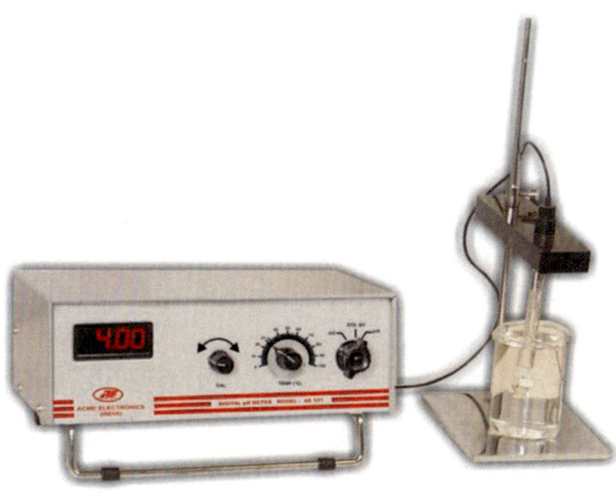

Figure 7.1. pH meter

PRINCIPLE

It is expressed in a number from 0 to 14. The number is an expression of the concentration of hydrogen ions (H^+) in the solution- the higher the concentration the more acidic the solution. The value is the negative exponent of the (H^+) concentration. A meter uses an ion-selective electrode (ISE) that ideally responds to only one specific ion, such as the H^+ concentration of the solution. When an ion diffuses from one region of activity to another, there is a free energy charge. The electrode produces a current, proportional to the concentration of H^+ ions, which is converted into a pH reading and displayed on the meter. Based on display type, pH meters are of two types: digital and analog.

A pH meter consists of two separate electrodes: glass and reference (or a single and combination electrodes), H^+ selective membrane and meter with control electronics and display. The round thin glass bulb at the bottom of a pH electrode, which forms the immersion tip in the solution, is the measuring part of the ph meter. The reference electrode is filled with the saturated KCl solutions that complete the measuring circuit and is not sensitive to any of the ions in the solution (Figure 7.2).

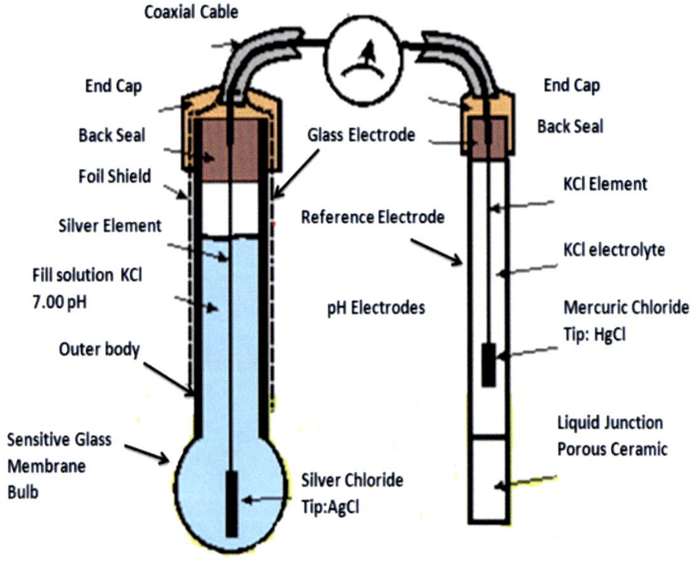

Figure 7.2 Construction of pH electrode

PROCEDURE

1. Calibration or standardization of pH meter
2. Select pH function by pressing the mode function.
3. Set the temperature with the temperature knob.
4. Remove the electrode from the storage solution rinse with distilled water and blot dry.
5. Immerse the electrode in pH 4 buffer and measure its pH.
6. Adjust display to read 4 with calibration knob.
7. Rinse the electrode and calibrate it with a second buffer of known vale (i.e 7 or 10 pH).
8. Measure the pH of the buffer (pH 4). The pH should read 4.
9. The pH meter is now calibrated and is ready to use for measuring the pH of any solution, by the procedure outlined below.
10. Remove electrodes from the buffer solution, rinse with distilled water and blot dry.
11. Place electrode in the test solution of an unknown pH.
12. Turn the function sector from the standby to pH
13. Read the pH of the test solution. Return function selector to stand by.

PRECAUTIONS

- Calibration should be done at the temperature of the test solution.
- Calibration should be preferably being done every day.
- The electrode should always be kept in distilled water, when not in use.
- Rinsing of electrodes with distilled water is a must before and after moving from one solution to another.

VIVA QUESTIONS

1. What is pH?
2. What is the meaning of acidic and alkaline?
3. What are the uses of pH meter?

4. What is the principle of pH meter?
5. What are three buffers used for calibrating a pH meter?
6. What is the function of KCl?

CHAPTER 8

WATER BATH

INTRODUCTION

A water bath is an apparatus fitted with a system for the control of temperature and hears the materials by moist heat. However, the use of such naked flame causes serious fire hazards in the laboratories, particularly when any flammable liquid is required heat. In the 1930s Glen and Ruth Morey resolved this problem by invented the heating mantle. The heating mantle is a reliable and non-flammable device used for heating with electric resistance wires woven into a fiberglass cloth sheath. Water baths are mainly of two types: constant temperature water bath and shaking water bath. Water baths are used for heating the solution and that material which cannot withstand dry heat (Figure 8.1).

These are used for melting of media, testing enzymatic activities, Widal test, Wassermann test for the cultures which require aeration as well as for the extraction of various plant extracts to be tested for pharmaceutical activities. Water baths have multiple usages in laboratories. In this water, the medium is used it have the ability to retain heat for a long time. Some of their usage includes (warming various chemicals and reagents, thawing of freeze samples; determination of coliforms, bacteriological examinations, cell cultivation, microbiological assays; corrosion tests; clinical laboratories; food and cosmetic industries and wastewater treatment industries.

Figure 8.1. Water bath

PRINCIPLE

It works on the principle of heating water with a heating element fitted with a thermostat for controlling the temperature and materials are heated with moist heat.

PROCEDURE

1. A constant temperature water bath is an insulated metallic vessel fitted with an electric heating mechanism and a thermostat for maintaining the water temperature.
2. There are racks for holding test tubes and flasks.
3. Water is filled in the container up to the top of the racks and heated at the desired temperature with the thermostat (Figure 8.2).
4. A glass beaker with water on a tripod and asbestos with a source of heat beneath (e.g Bunsen burner) can be used as a simple water bath in the laboratory.
5. The shaking water bath with a shaker is a thermostatically controlled apparatus. It is used for the cultivation of microorganisms in a broth medium.

6. It is better than an incubator because it provides a rapid and uniform transfer of heat to the culture vessel and its agitation provides increased aeration, resulting in acceleration of growth and production of enzymes.

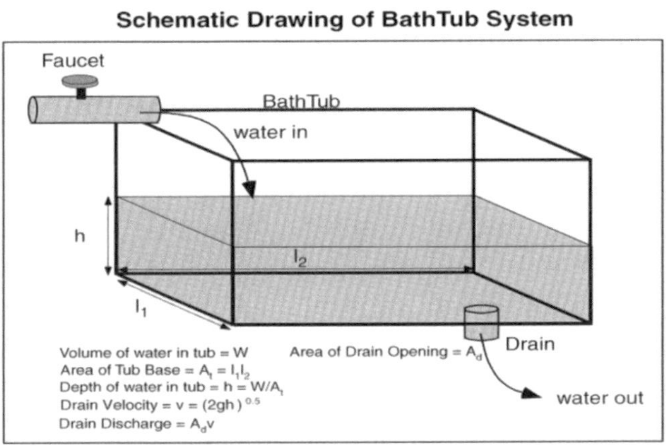

Figure 8.2. Schematic diagram of a water bath

PRECAUTION

- Always maintain the proper water levels in the bath.

VIVA QUESTIONS

1. What is a shaking water bath?
2. What are its uses?
3. What are the uses of a water bath?
4. What is the principle of a water bath?
5. How many types of water bath?

CHAPTER 9

AUTOCLAVE

INTRODUCTION

An autoclave is a sealed heating device, which uses moist heat under pressure essentially a double jacket steam chamber equipped with devices that permit the chamber to be filled with saturated steam and maintained at a designed temperature and pressure for any period of time. It was developed by Charles Chamberland, one of the Pasteurs associative in 1884. It is a common instrument for very effective sterilization. The pressure increases the boiling water can be sterilized. Steam at a pressure of 15 psi (121°C) kills all organisms. It is the increased temperature and not the increased pressure that kills microorganisms in this procedure. It is used for sterilizing culture media, heat-stable liquids, utensils, linens, bandages, hospital and laboratory equipment not damaged by heat and moisture. According to our ancient history a scientist named Moses, who wrote that the purification of chemicals was achieved by using boiling water and fire. In 460-377 BCE, Hippocrates suggests that the sterilization of surgical instruments can be done by successfully pouring boiling water onto them. After some time in 1729, a famous scientist Lazzaro Spallanzani proved that sterilization of instruments requires only 30 minutes by heating them in a sealed glass (Figure 9.1).

PRINCIPLE

It is based on the principle where moist heat under pressure at elevated temperature above 100°C effects sterilization. The higher pressure in the autoclave, than higher the temperature of an autoclave. Saturated steam heats an object above 2500 times more efficiently than does the hot air at the same temperature. Moist heat kills microbes by coagulating proteins/enzymes (denaturation), which is caused by the breakage of the hydrogen bonds. This coagulation process is familiar to anyone who has watched an egg white frying.

A simple autoclave is a double- walled cylindrical metallic vessel, made of thick stainless steel or copper, one end of which is open to receive the material (Figure 9.2). The lid of the autoclave is provided with the pressure gauge, steam cock (exhaust valve) controls for pressure adjustment, and safety valve. Basket provides with holes all around, for the free circulation of the steam is used for keeping the materials. Metallic separations, having holes, are used if more materials are to be sterilized. There are horizontally autoclaved in which pressure, temperature and steam supply are automatically controlled. The steam is heated in the jacket of an autoclave, enters the sterilization chamber through an opening at the upper rear and is expelled through a vent at the bottom front (Figure 9.3).

Figure 9.1. Autoclave

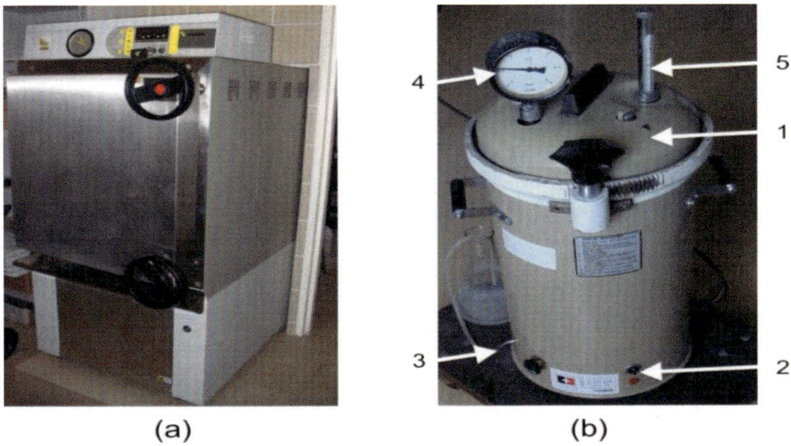

Figure 9.2. Sterilisation by heat the autoclave. (a) Bigger, Automatic Autoclave Operated by External Steam Afflux. (b) Smaller, Manual Autoclave: 1. Lid 2. Power Switch 3. Bleeder Valve 4. Pressure Gauge 5. Thermometer.

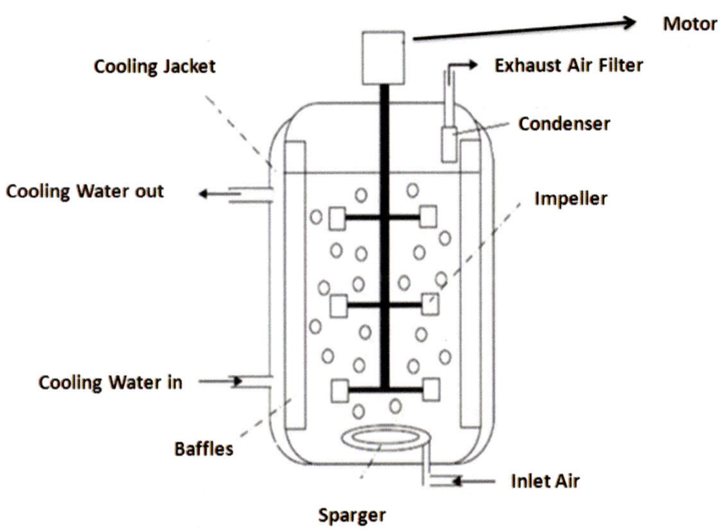

Figure 9.3. A line image of autoclave

PROCEDURE

1. Place the items to be sterilized in the chamber.
2. Place the pressurized lid, close and lock the autoclave door by tightening the screw fastener diagonally.
3. Switch on the power supply.
4. When all the air has been ejected, the temperature reaches 100°C close the operating valve.
5. Check the jacket pressure.
6. Note the time from the moment temperature reaches 121°C/15lbs pressure.
7. Sterilize for 25-20 minutes.
8. Switch of the electric supply.
9. Allow the autoclave to cool till the pressure falls to the zero mark.
10. Unlock and open the autoclave lid.
11. Remove autoclave items from the chamber.

PRECAUTIONS

- The level of water must be checked before operating.
- Articles should be placed loosely so the steam can easily penetrate them.
- Air should be expelled.
- Wrapping objects in aluminum foil is not recommended for dry materials.
- Never open the lid unless the pressure of the chamber comes down.
- Wear insulated gloves when handling hot autoclaved items.
- Cotton plugs should be tightly fitted.

VIVA QUESTIONS

1. What is an autoclave?
2. What is the principle of autoclave?
3. What items are not be autoclaved and why?

4. What indicator organism is used to test the efficacy of an autoclave?
5. What organisms are not killed in an autoclave at 121°C?
6. What size of autoclave do we need in the laboratory?
7. What kind of loads will be running in the autoclave?
8. What is the steam source for the autoclave?
9. What conditions are required to prepare for sterilization?
10. How many years have the autoclave manufacture in the market?

CHAPTER 10

BUNSEN BURNER

INTRODUCTION

The Bunsen burner name discovered by R.W. Bunsen is a type of gas burner with which a very hot, practically non-luminous flame, the temperature of which reaches 1870°C at its hottest point, is obtained by allowing the air to enter at the base and mixes with the gas. Direct flaming is one of the simplest methods of dry heat sterilization. It is used in the microbiology laboratory for sterilizing inoculating loops/needles, mouths of culture tubes by incineration and to dry the inside of pipettes (Figure 10.1). Incineration (sterilization procedure by which contaminants are burned to ashes) in a flame is perhaps the most rigorous of all heat treatments for sterilization.

Figure 10.1. Bunsen burner

Aseptic transfer of a culture from a culture vessel to another is successful only if no contaminating microorganisms are introduced in the process. The transport/transfer of organisms is performed with an inoculating loop or needle. To effectively sterilize, the inoculating loop, prior to picking up the organisms, the end of the wire and the entire loop must be held in the Bunsen burner flame until they glow red hot since the loop is not sterilized by simple moving over a Bunsen burner. The Bunsen burner is provided with the barrel, collar, gastap, air hole, gas jel and base (Figure 10.2)

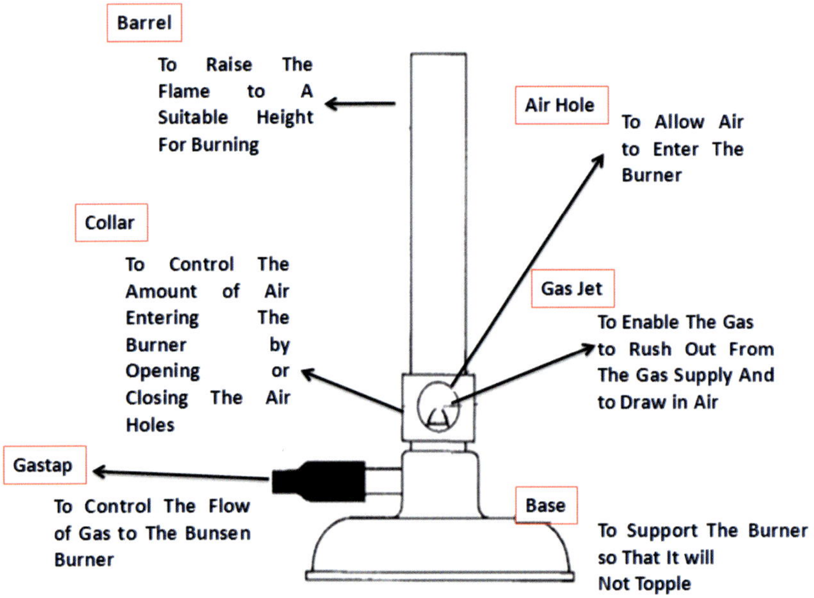

Figure 10.2. Heat the Flame: Barrel, Collar, Gastap, Air Hole, Gas Gel and Base

PRINCIPLE

The temperature reaches 1870°C at its hottest point, which is obtained by allowing the air to enter at the base and mixes with the gas. Direct flaming is one of the simplest methods of dry heat sterilization.

PROCEDURE

1. Open the desk outlet value.
2. Connect Bunsen Burner to gas supply.
3. Turn on the gas valve and provide light to the Bunsen burner.
4. Adjust the flame, so the blue flame generates on the inner blue cone.

PRECAUTIONS

- Do not allow the formation of floating ashes and aerosols to avoid the spreading of infectious agents.
- Do not go close when you hold alcohol in your hand.

VIVA QUESTIONS

1. What is the use of Bunsen burner?
2. What is the principle of Bunsen burner?
3. What is the use of Bunsen burner in the microbiology laboratory?
4. How are the microorganisms are killed in the open flame?

CHAPTER 11

INCUBATOR

INTRODUCTION

The incubator is an apparatus in which media inoculated with microorganisms are cultivated at a constant temperature favorable to the growth of these microorganisms. The very first incubators discovered in Egypt and China: these were used in the hatching of fertilized chicken eggs, they consist of fire-heated rooms in which eggs were placed. An incubator similar to a hot air oven consists of an insulted, two walled cabinet, fitted with a heating element at the bottom. It has a thermostat to control the temperature (Figure 11.1). The chamber is provided with perforated shelves adjustable to a convenient height for keeping the items inside.

Figure 11.1. Incubator

The incubators are provided with double doors, the inner one made of glass so that the contents and microbial growth in the incubator could be viewed without making a change in the temperature of incubation and growth of the organisms. Both low and high temperatures, then the room temperature, can be maintained in an incubator, depending upon the nature of the microorganism being cultured. Incubators are available in which CO_2/O_2 concentration can be maintained (Fig 3.5.1). This device helps to maintain optimal humidity, temperature and other gaseous conditions such as oxygen (O_2) and carbon dioxide (CO_2). It is used in microbiology, biotechnology and pharmaceutical industries for culturing microbes at the desired temperature and oxygen requirement.

PRINCIPLE

It is based on the principle of providing a constant temperature provided through hot-air/dry heat for maintaining temperature above the room temperature and cool air produced through a cooling system for maintaining temperature below the room temperature. The temperature of an incubator is kept constant by thermostat control for culturing microorganisms at the desired temperature; the variation of temperature should not be more than one degree Celsius (1°C). While culturing thermophilic microorganisms or incubating cultures for a longer duration, beaker-filled water (sterilized water) is kept in the incubator to provide moisture or the agar plates are placed in the moist chamber with damp sterile cotton wool at the bottom.

PROCEDURE

1. Firstly start the incubator by providing a power supply.
2. Switch "on" of main in the lab.
3. Turn on the red color power knob and cooling knob towards 0-1.
4. Set temperature of the incubator with the help of the driver.
5. If the temperature is not shown, check the power supply button.
6. The simplest incubators are insulated boxes with an adjustable heater typically going up to 60°C to 65°C (140°F to 150°F), though some can go slightly higher (generally not more than100°C).
7. The most commonly used temperature both for bacteria such as the frequently used *E. coli* as well as for mammalian cells is

approximately 37°C, as these organisms grow well under such conditions.

8. For other organisms used in biological experiments, such as the budding yeast *Saccharomyces cerevisiae*, a growth temperature of 30°C is optimal (for general fungi the optimal temperature is 27°C).

PRECAUTIONS

- The incubator door should be open only when required.
- It should be taken for not allowing the agar media to dry due to excessive evaporation.
- Simple incubators without a cooling system should be avoided for culturing microbes in the summer season especially in tropical countries.

VIVA QUESTIONS

1. What is the principle of an incubator?
2. What is the difference between an incubator and a hot air oven?
3. What is the important precaution taken for culturing thermophilic microbes in an incubator?
4. What is the use of an incubator and how many types of an incubator?

CHAPTER 12

HEMACYTOMETER

INTRODUCTION

Hemocytometer is a device used to count cells. It was originally designed for the counting of blood cells. The hemocytometer was invented by Louis-Charles Malassez and consists of a thick glass microscope slide with a rectangular indentation that creates a chamber. This chamber is fixed with a laser-etched grid of perpendicular lines (Figure 12.1). The device is carefully crafted so that the area bounded by the lines is known, and the depth of the chamber is also known. It is therefore possible to count the number of cells or particles in a specific volume of fluid and thereby calculate the concentration of cells in the fluid overall.

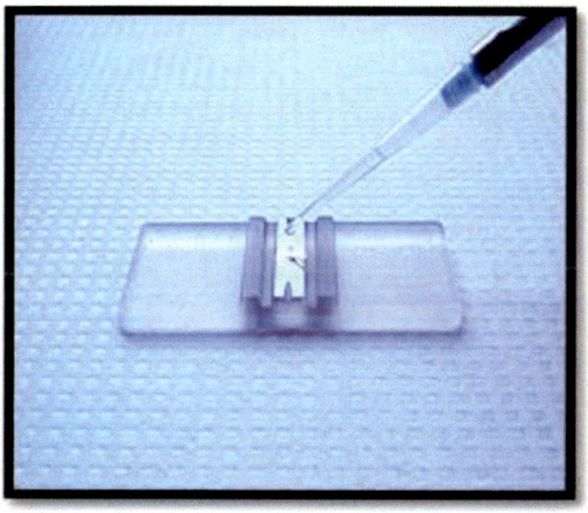

Figure 12.1. Hemacytometer

PRINCIPLE

The gridded area of the hemocytometer consists of nine 1 x 1 mm (1 mm^2) squares. These are subdivided in 3 directions; 0.25 x 0.25 mm (0.0625 mm^2), 0.25 x 0.20 mm (0.05 mm^2) and 0.20 x 0.20 mm (0.04 mm^2). The central square is further subdivided into 0.05 x 0.05 mm (0.0025 mm2) squares. The raised edges of the hemocytometer hold the coverslip 0.1 mm off the marked grid, giving each square a defined volume (Figure 12.2).

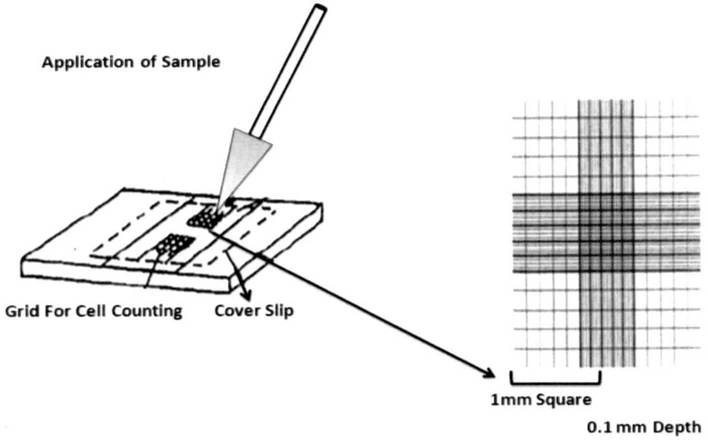

Figure 12.2. Hemocytometer Lid

Dimensions	Area	Volume at 0.1 mm depth
1 x 1 mm	1 mm^2	100 NL
0.25 x 0.25 mm	0.0625 mm^2	6.25 NL
0.25 x 0.20 mm	0.05 mm^2	5 NL
0.20 x 0.20 mm	0.04 mm^2	4 NL
0.05 x 0.05 mm	0.0025 mm^2	0.25 NL

PROCEDURE

1. Take a suspension of cells
2. Prepare a 1:1 dilution of the suspension of cells in trypan blue.

3. Load the sample in a hemocytometer.
4. The cover slip is carefully put on the sample side. Overloading the sample should be avoided.
5. Determine the number of cells (total& viable).
6. View the cell under the microscope at 100X magnification.
7. Calculate the cell no. 4 corner squares and also the average
8. Clean the hemocytometer after use.

PRECAUTIONS

- To prepare the counting chamber, the mirror-like polished surface is carefully cleaned with 75% ethanol.
- The use of a mechanical suction device to draw up fluids into the pipette.
- Blood and the diluting fluid should be properly mixed.
- Do not allow overflowing of the cell suspension at the time of filling of the cambers.
- The coverslip is also cleaned.
- They must be heavy enough to overcome the surface tension of a drop of liquid.
- A coverslip is placed on the counting surface prior to putting on the cell suspension.
- Introduced any of the cell suspension into any of the V-shaped wells with a micropipette.
- The area under the coverslip fills by capillary action.
- Sufficient liquid should be introduced so that the mirrored surface is just covered.
- The charged counting chamber is placed under the microscope stage and the counting grid is brought into focus at low power.
- Repeat count to check the accuracy.

VIVA QUESTIONS

1. What is a hemacytometer?
2. What is the number of total quadrants in a chamber?
3. What is the surface area of a counting chamber?

CHAPTER 13

HOT AIR OVEN

INTRODUCTION

A hot air oven consists of an insulated cabinet which is held at a constant temperature by means of an electric heating mechanism and thermostat. A hot air oven is used to sterilize glassware, needles, glass syringes, and metal objects and is the only means of sterilizing oils, waxes and powders. It is a very effective method of sterilization requiring 180°C for 20 minutes, 171°C for one hour, and 160°C for 2 hours or 120°C for 16 hours or longer, depending on the volume (Figure 13.1).

Figure 13.1. Hot Air Oven

PRINCIPLE

Sterilization is accomplished by hot-air, a kind of dry heat sterilization. Dry heat penetrates substances more slowly than steam heat. Dry heat is microbicidal and kills the microbes by oxidation of molecules.

A hot air oven consists of an insulated cabinet, fitted with a thermostat to control the temperature, a fan to keep the hot air circulating at a constant temperature, a thermometer, and perforated shelves for keeping the materials. The heating in an oven is through the coils fitted at the bottom of the cabinet.

PROCEDURE

- Place the items to be sterilized on the shelves.
- Close the door tightly.
- Switch on the power supply.
- Noted time from temperature reaches 120°C or 170°C.
- Sterilize for the desired time period.
- Switch off the power supply.
- Allow to cool.
- Items will be sterilized and ready to use.

PRECAUTIONS

- Glassware items should be dried before putting them in the oven.
- Never open the door when the temperature comes down to room temperature.
- To avoid breakage of the glassware items.
- Paper wrapped items are not to be kept in the oven for sterilization.

VIVA QUESTIONS

1. Explain how dry heat kills microbes?
2. What items are sterilized in the oven?

CHAPTER 14

VACUUM FILTRATION ASSEMBLY

INTRODUCTION

Filtration is the passage of material, liquid or gas through a filter or straining device. It can be used to sterilize substances that are destroyed by heat such as vaccines, some culture media, and enzymes, antibiotic solutions to separate viruses and to collect microorganisms from air and water samples (Figure 14.1).

Figure 14.1. Vacuum filtration assembly

- **Glass Microfiltration**: Support Systems

PRINCIPLE

Vacuum filtration is a technique for separating a solid product from a solvent, a liquid reaction mixture, or microbes from a substrate which is poured through a filter paper. The filter traps the solid/microbes and the sterile liquid is drawn through the funnel into the flask below by vacuum. The apparatus used for this purpose is called a filtration assembly. A vacuum is created in the receiving flask; air pressure then forces the liquid through the filter. Membrane filters of different pore sizes ranging from ultrafilters (small molecules), 0.025 μm (smallest viruses, molecules), and 0.45 μm (a few bacteria, viruses, molecules) to 10 μm are used depending upon the particles size.

REQUIREMENTS

- Filtration assembly
- Filter flask
- Filter papers
- Membrane filter disks
- Disposable dropping bottle
- Vacuum pump
- Sample

PROCEDURE

1. Turn on the vacuum.
2. Check the vacuum by feeling for suction at the end of the tubing.
3. Attach the tubing to the sidearm of the filter flask and check the suction at the top of the flask. Put the black rubber ring adapter on the top of the flask and then the Buchner funnel.
4. Proper suction to be checked again.
5. Place a filter paper/ membrane filter in the Buchner funnel.
6. Slowly pour the sample into the funnel.
7. Continue to draw air (vacuum) through the sample till it is completed filtered.

8. Disconnect the vacuum source.
9. Disassemble the membrane filtering unit.
10. Decant the filtered sample from the filter flask.

PRECAUTIONS

- A cotton plug should be used in the vacuum line to ensure sterility
- Always use a sterile filter flask
- Sterile membrane filters should be used.

VIVA QUESTIONS

1. What is the principle of filtration?
2. What are the advantages of filtration?
3. Where is membrane filtration is used in the microbiology laboratory?
4. What substance should be sterilized by filtration?

CHAPTER 15

SPECTROPHOTOMETER

INTRODUCTION

Most of the biochemical experiments involve the measurement of a compound or group of compounds present in a complex mixture. Colorimetry and spectrophotometry are the most widely used methods for determining the concentration of chemical compounds. The basic arrangement of a typical colorimeter is shown in (Figure 15.1).

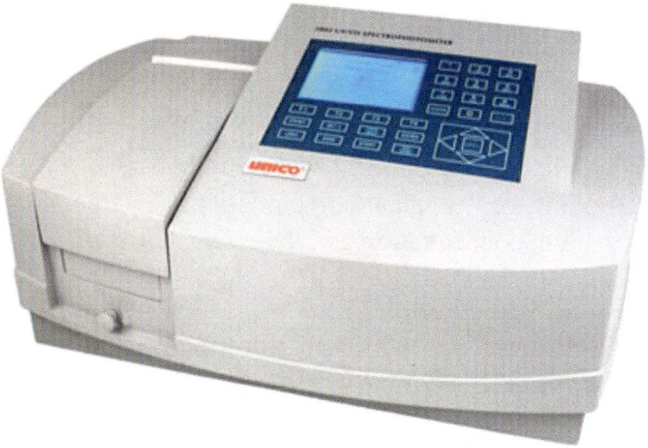

Figure 15.1. Spectrophotometer

In 1935, Arnold O. Beckman started working with National Technologies Laboratories, later this company named s Beckman Instruments. In 1940, the first spectrophotometer "Beckman DU spectrophotometer" was sold. It is a device that is commonly used for the measurement of light intensity generated by either reflectance or transmittance of particular substances (solutions, gasses and opaque or transparent). They work in the 200 nm-2500 nm wavelength range of light using different calibrations and controls. Machine calibrations are needed within these ranges of light, by using standards that are of different types depending on the wavelength of the photometric determination.

PRINCIPLE

A spectrophotometer measures the transmission or absorption of light in liquids as a function of wavelength. The amount of light absorbed is proportional to the intensity of the color and hence to the concentration.

- The amount of light absorbed is proportional to the intensity of the color and hence to the concentration. There are two laws on which the principle of colorimetry is based:
- Lambert's Law-when a ray of monochromatic light passes through an absorbing medium, its intensity decreases exponentially as the length of the absorbing medium increases.
- Beer's law- When a ray of monochromatic light passes through an absorbing medium, its intensity decreases exponentially as the concentrations of the absorbing medium increases. These two laws combined to call Lambert-Beer law.
- When light of a specific wavelength passes through a liquid sample there is a quantitative relationship (Beer's law) between the solute concentration and the intensity of the transmitted light (Figure 15.2), that is,

$$I=I_0 * 10^{-kcl}$$

- where I sub 0 is the intensity of transmitted light using the pure solvent, I is the intensity of the transmitted light when the colored compound is added, c is the concentration of the colored compound, l is the distance the light passes through the solution, and k is a constant.

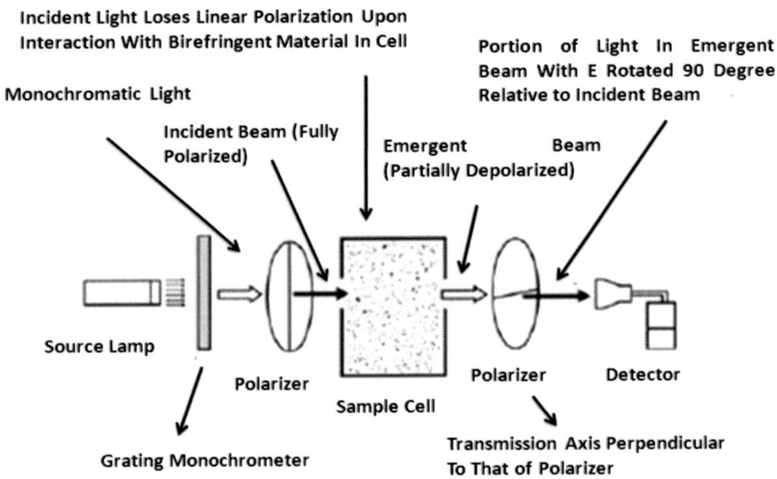

Figure 15.2. Schematic diagram of a modified spectrometer

A spectrophotometer consists of two parts, *spectron* = for produce light of any selected color (wavelength) and photometer for measuring the intensity of light. In the spectrophotometer, a beam of light is transmitted through a solution or bacterial suspension to a light-sensitive detector (Figure 15.3).

Concentration is measured by the extent of absorption of light at the appropriate wavelength registered on the instrument's scale as the percentage of transmission. Also printed on the instrument scale is a logarithmic expression called the absorbance (also called as optical density or OD) which is calculated as Abs = 2 – log of % transmittance. The absorbance is used to plot bacterial growth. For determining the number of bacteria through turbidity, the bacterial suspension could have more than 10 million cells/ml. Since more than a million cells/ml must be present for the first traces of turbidity to be visible. The higher absorbance limit; the greater the concentration of bacterial cells.

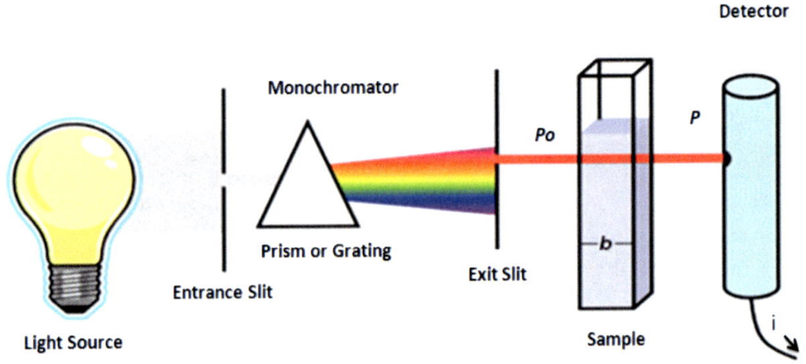

Figure 15.3. UV-VIS spectrometer

PROCEDURE

1. Rinse the cuvette several times with distilled water.
2. Switch on the spectrophotometer, 30 minutes before its calibration.
3. Rotate the zero control knobs clockwise.
4. Set the desired wavelength knob e.g 686 nm for measuring turbidity.
5. Adjust the meter needle to zero by rotating the zero control knob.
6. Pour 3 ml sterile nutrient broth into the well-rinsed cuvette.
7. Open the cover and insert the cuvette into the sample holder.
8. Close the cover.
9. Keep the index line of the cuvette in line with the index line on the sample holder.
10. Rotate the light control knob to adjust the meter to reach 100% transmittance.
11. Remove cuvette of nutrient broth and close lid.
12. Read, if the needle does not return to zero, readjust accordingly.
13. Reinsert nutrient broth cuvette again to see if 100% transmittance still registers.
14. Transmittance between 0 and 100% shows the calibrations of the instrument and ready for turbidity measurements.

PRECAUTIONS

- Always switch on the instrument 30 minutes before use.
- Don't wipe the cuvette with a towel.
- Any make or fingerprints are to be avoided on cuvettes as these will hinder transmittance. Always reset the instrument before using it again for the next sample.
- Handle the cuvettes with great care in order to avoid their breakage.

VIVA QUESTIONS

1. What is the principle of a spectrophotometer?
2. What are the uses of a spectrophotometer?
3. What is the advantage and disadvantage of estimating bacterial population by measuring turbidity?
4. What is Lambert's Law and Beer's Law?

CHAPTER 16

LAMINAR AIR FLOW

INTRODUCTION

Laminar airflow (LAF) bench or hood is equipment equipped with a filtered air supply that is free from the microorganisms for use in microbiology and tissue culture laboratories, as well as in pharmaceutical industries. It is provided with a special filter system that uses high-efficiency particulate air (HEPA) filters, which are made of boron silicates fibers arranged into a flat sheet.

A laminar airflow cabinet (Figure 16.1) is a closed closet that provides a contamination-free environment. In the hood, the air is drawn through a HEPA (high-efficiency particulate air) filters and blown in a very smooth, laminar flow towards us. This hood is usually made up of stainless steel with a smooth surface and there are no gaps or joints where spores might collect. These can remove almost all microorganisms larger than 0.3 μm in diameter from the air entering the bench area (Figure 16.1).

Figure 16.1. Laminar air flow cabinet

Even though viruses are as small as 30 nm when air-borne they are usually attached to larger particles of dust. It is used for performing the pouring, plating, inoculating, streaking, without any kind of contamination. In the 1960s, the United States' biggest political enemy was the Soviet Union. But in the race to build deadlier nuclear weapons, sometimes the enemy was smaller than one-tenth the size of a human hair. In the immediate post-war era, Sandia focused primarily on weapons development, pushing the limits on miniaturization in both mechanical parts and transistors.

PRINCIPLE

It works on the principle of air sterilization by filtration. In the laminar flow, a dust-free air current flows with uniform and constant velocity along parallel lines in a single direction which helps in creating aseptic conditions in the working area. UV light fitted in the chamber kills the air-borne microbes present in the chamber. The principle of a laminar air flow hood is to achieve contamination-free environment and protect the work from cross-contamination. Contaminated air enters from inside the front wall to the hood via HEPA filters (filter size 0.3 μm) to the clean bench (Figure 16.2). HEPA filters are considered as the heart of all laminar airflow hoods. This filter removes small particulates generally they are airborne, such as microorganisms. However, it does not provide filtration to vapors or gasses.

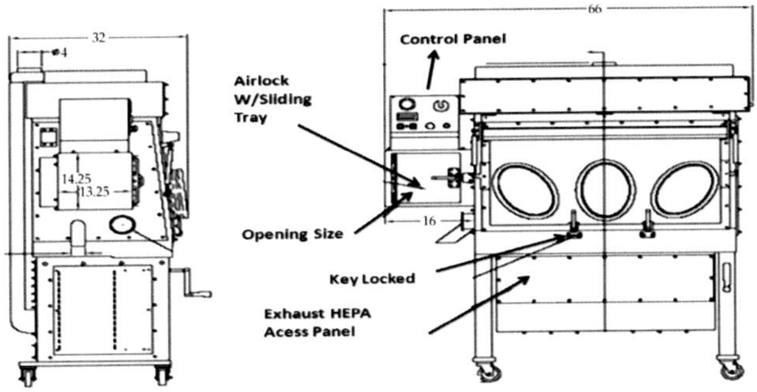

Figure 16.2 Schematic line diagram of laminar air flow (control panel, Key locked, exhaust HEPA access panel, opening size and airlock sliding tray).

PROCEDURE

1. Switch on the UV light for 30 minutes with the font covering glass still closed, before starting an experiment.
2. The covering glass of the LAF bench is opened; the air blower is set at the desired degrees to expel the contaminated air still present in the chamber.
3. Now sit properly in front of the chamber.
4. Wash hands and also the stage of the chamber with alcohol.
5. Place the required materials side by side on the left-hand side of the stage and now perform an experiment within the flame zone of the Bunsen burner/spirit lamp.
6. After the experiment is over, stop the motor, wash again the working table with alcohol.

PRECAUTIONS

- We must wash our hands with detergent.
- We should not talk inside the chamber.
- UV light should never be on while doing an experiment.
- Cleaning of filters at intervals is to be done.
- Avoid using air supply when you are working.

VIVA QUESTIONS

1. What is the principle of a laminar air flow bench?
2. What types of filters are used in the LAF bench?
3. What is the size of the pores of these filters?
4. How has the LAF bench been named?
5. What precautions are needed while using the LAF bench?

CHAPTER 17

MICROSCOPE

INTRODUCTION

Microorganisms are too small to be seen with the unaided eye. The study of microorganisms is greatly enhanced by the use of microscopes. The word microscope, first coined by Giovanni in 1625, derived from the Latin word micro (small) and the greek word scopos (to look at), is an important tool for microbiologists and the technology of making very small things visible to the human eye is called microscopy (Figure 17.1). The simple microscope used by Anton van Leeuwenhoek in the 17th century in 1673 had only one lens and was similar to a magnifying glass that could magnify a microbe 300X.

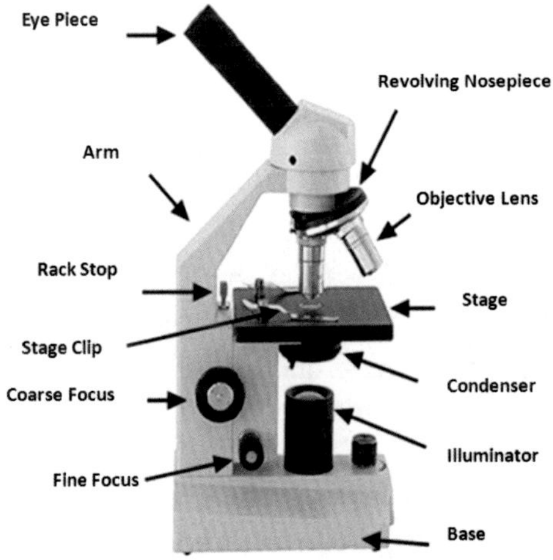

Figure 17.1. Microscope

The simple microscopes enabled him to be the first person to see bacteria and protozoa. Robert Hooke, a contemporart of Leeuwenhoek built compound microscopes that have multiple lenses. The microscopes used today are dividing into two categories: light and electron. Light microscopes are sub-divide into bright-field and fluorescent microscopes, while electron microscope is subdivided into scanning and transmission electron microscopes.

The bright-field microscope or compound light microscope that allows light rays to pass directly through to the eye without being deflected by an intervening opaque plate in the condenser is routinely used in all the microbiology laboratory to study the size, shape, arrangement of microbial cells and sporulating structures.

Dark-field microscopy is a kind of light microscopy that uses a special dark-field condenser that contains an opaque disk that causes the light to reflect off the specimen at an angle rather than pass directly through it. This microscopy is used to examine live microorganisms. Fluorescent microscopy uses ultraviolet light instead of white light to excite molecules within the specimen or a fluorescent dye. Fluorescence microscopy must be intrinsically fluorescent or stained with a dye such as auramine O, acridine orange, fluorescein, DAPI (diamidino-2-phenylinodole).

BRIGHT-FIELD LIGHT MICROSCOPY

A microscope is a compound optical system, a compound magnifying glass. The real meaning of the functioning of a microscope is test object positioned between the single and double focus points of the objective. The light coming from the object and passing through the lens creates a magnified, inverted and real image of the subject on the other side of the objective, behind the double focus point.

The eyepiece is at a distance from the objective lens so that the image formed by the objective is generated within the focus of the eyepiece. Thus, looking throughout the eyepieces, one can see a further enlarged, direct but virtual image of this real, inverted and magnified image. The magnification power of a bright-field microscope can be calculated by multiplying the magnification of the objective and of the eyepiece, respectively. The objective lens system consists of multiple lenses. The first member is the front lens facing the object. This determines the magnification and resolution of the microscope. Other elements are responsible for the exclusion of lens errors.

The features of an image formed by the objective depending on the optical characteristics of the objective lens.

The quality of a lens represents its property of how sharp the object image is drawn. The image defect, i.e. blurring is caused by lens errors (aberrations). The spherical aberration (spherical divergence) is caused when the lens away from the optical axis increasingly breaks the light rays passing through it, therefore a point-like object in the image will be blurred at the edge. The reason for chromatic aberration is that the focal points of different wavelengths of light do not coincide on the optical axis. The shorter wavelength rays unite closer, while longer-wavelength rays unite farther. Thus, a sharp image cannot be obtained using white light, and the image has a colored (rainbow) border. A concave flint glass lens with lead content and high refractive index is fitted to the convex lens to correct for chromatic aberration (the achromatic lens has two-color correction, while the apochromatic lens has three-colour correction).

The resolution of the objective lens is the ability of how detailed the image of a subject can be drawn. The resolving power is quantified with the minimum distance between two points that are just distinguishable. The resolution (d) depends on the illumination wavelength of light used (λ), the half-angular aperture of the objective lens (α) and the refractive index of the material between the front lens and the coverslip (n).

$$d = 1.22\,\lambda\ /2\ n \times \sin\alpha$$

where $n \times \sin\alpha$ = numerical aperture (NA).

The greater the resolution, the smaller the value of d is. This can be achieved by reducing the wavelength of light used, increasing the angular aperture of the objective, or increasing the refractive index of the material between the front lens and the coverslip. Using a light microscope, there is an opportunity only to change the latter. For this purpose, cedar oil (oil immersion) is the most suitable material, because it has almost the same refractive index as that of glass, so the light passes through a virtually homogeneous medium. The denominator of the above formula, $2\ nX \sin\alpha$ is the value of numerical aperture (NA), which may vary between 0.20 and 1.4 (always indicated on the lens). The eyepiece draws a direct image of the test object. The fine structure of an image observed in the microscope depends on the details of the real picture, which in turn is determined by the resolution of the objective. This image, however, is not visible to the naked eye; it can only be visualized in the magnification of the eyepiece.

PRINCIPLE

A microscope is a series of finely ground lenses and visible light as a source of illumination are used. The microscope field or area observed is brightly lighted and the objects being studied appear dark because they absorb some of the light. However, microbes absorb light after staining them with a dye greatly increases their light-absorbing ability resulting in greater contrast and color differentiation.

REQUIREMENTS

- Environmental water sample (e.g. from a lake, a stream, a creek or an aquarium)
- Glass slides
- Cover slips
- Glass dropper dispenser alcohol (for sterilization)
- Bunsen burner
- Light microscope

PROCEDURE

- Dry the surface of a glass slide with alcohol over a Bunsen burner and then label the slide
- Place one drop of the sample to the slide with a glass dropper.
- Put the edge of a cover slip on the slide so that it touches the edge of the water drop.
- Slowly lower the cover slip to prevent the formation and trapping of air bubbles.
- Put the sample under the microscope, and locate the focal plane.
-
- Apply first an objective lens of 16X, and then 40X magnification.
- Observe the shape and movement of microbes, in the case of protists and eukaryotic algae, try to identify the different cell organelles (e.g. chloroplasts, contractile vacuoles) Figure 17.2.
- Make drawings about the observations.

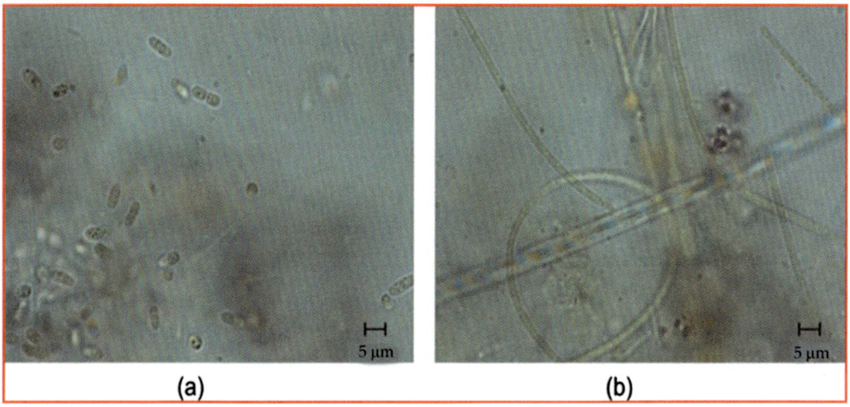

Figure 17.2. Bright-field micrograph of microorganisms from natural waters (a) rod and (b) filamentous bacteria from an artificial pond.

FLUORESCENCE MICROSCOPY

Specimens that absorb light of color and subsequently emit light of another color (fluoresce) can be visualized by using the fluorescence microscope. The basis of fluorescence microscopy is the principle of the removal of incident illumination by selective absorption, where light absorbed by the specimen and re-emitted at another wavelength is transmitted. The light source must produce light of appropriate wavelength and ineffective wavelengths that are unable to excite the fluorochrome used are removed by an excitation filter. The phenomenon when a sample (e.g. cell organelle of a microbe) shows fluorescence without prior staining is called auto-fluorescence. Otherwise, bacterial cells can be stained with a fluorescent dye. Using a fluorescent dye, e.g. DAPI (4',6-diamidino-2-phenylindole), cells are stained bright blue because the dye forms a complex with the cell's DNA. Fluorescent dyes are therefore widely applied to visualize and enumerate bacteria in different natural habitats or clinical samples.

Microbes Examine with DAPI Staining (Microbes from Water Sample)

DAPI molecules are able to penetrate cell membranes and bind to the double helix of the DNA. Cells can be easily counted, if a known volume of the fixed water sample is filtered through a membrane and, after staining the filter, the surface is investigated by a fluorescence microscope (Figure 17.3).

PRINCIPLE

A microscope is a series of finely ground lenses and visible light as a source of illumination is used. The microscope field or area observed is brightly lighted and the objects being studied appear dark because they absorb some of the light. However, microbes do not absorb much light but staining them with a dye greatly increases their light-absorbing ability resulting in greater contrast and color differentiation. The effective magnification of a microscope depends upon its resolving power.

- Typical magnification values are:
- Scanning 10X × 4X= 40X magnification
- Low power 10X × 10X = 100X magnification
- High power 10X × 40X = 400X magnification
- Oil immersion 10X ×100X = 1000X magnification
- Total magnification = Magnifying power of the objective lens* magnifying power of the ocular lens.
- The major parts of a microscope and their function are shown in Fig 17.1.

REQUIREMENTS

- Disposable gloves laboratory scales paraformaldehyde (PFA) Beaker
- Magnetic stirrer
- Phosphate buffered saline (PBS) NaOH solution
- Pasteur pipette
- Membrane filtration apparatus
- Polycarbonate and cellulose nitrate membrane filters (0.22 or 0.45 µm pore size) 50 ml Falcon tube
- Plastic Petri dishes scalpel
- Pipette with pipette tips DAPI solution (1µg/ml)
- 80% ethanol
- Double distilled water glass slide
- Cover slip
- Vectashield Mounting Medium (H-1000, Vector Laboratories Ltd) immersion oil (non-fluorescent)

- Epifluorescence microscope with a mercury lamp digital camera computer with software

PROCEDURE

1. For the preparation of the fixative solution, dissolve 1 gm PFA in 50 ml PBS. (PFA causes irritation when inhaled; therefore the use of a fume hood is recommended.)
2. Dissolution can be aided with heating (60°C), permanent stirring and adding some drops of NaOH solution.
3. Adjust pH to 7.0.
4. Filter the solution through a 0.22 μm pore size membrane filter. (The prepared 2% PFA solution can be stored in the fridge for one week).
5. Filter the water sample (2-50 ml, depending on the type of sample) using a polycarbonate membrane filter (slowly with occasional stirring).
6. To help uniform cell distribution, place a 0.45 μm pore size cellulose nitrate membrane filter between the sieve of the filtration unit and the polycarbonate filter.
7. Fill up the Falcon tube with fixative (PFA solution) and immerse the filter in it with sterile forceps (PFA solution must cover the entire membrane filter).
8. Incubate the filter overnight at 4°C.
9. Fill PBS into an empty Petri dish, then transfer the filter into the PBS solution for 1-2 minutes (liquid must cover the entire membrane filter).
10. Transfer the filter to another empty Petri dish and let it dry.
11. Cut a 0.5 by 0.5 cm piece from the filter with a scalpel or scissors, and pipette 30 μl DAPI solutions onto its surface. From this step onwards, work in a dark place.
12. The filter piece can be marked with a soft pencil.
13. After 2 minutes, transfer the filter into 80% ethanol for a few seconds.
14. Dip the filter paper into double distilled water for a few seconds.
15. Dry the filter.
16. Place the filter onto the surface of a glass slide, put a drop of Vectashield Mounting Medium onto the filter and then cover with a cover slip.
17. Cover with a paper towel and gently press the cover slip to remove any excess of mounting medium.

18. Examine the slide with an epifluorescence microscope using a 100X objective and immersion oil under UV excitation (the absorption maximum of DAPI is at 358 nm, and the emission maximum is at 461 nm).
19. Record images from at least 20 different microscopic fields with a digital camera.
20. Count the cells on each picture and determine the mean values.
21. Determine the cell count for one ml of water sample based on the amount of filtered water and the size of the membrane filter field. Evaluate the variability of cell morphology.

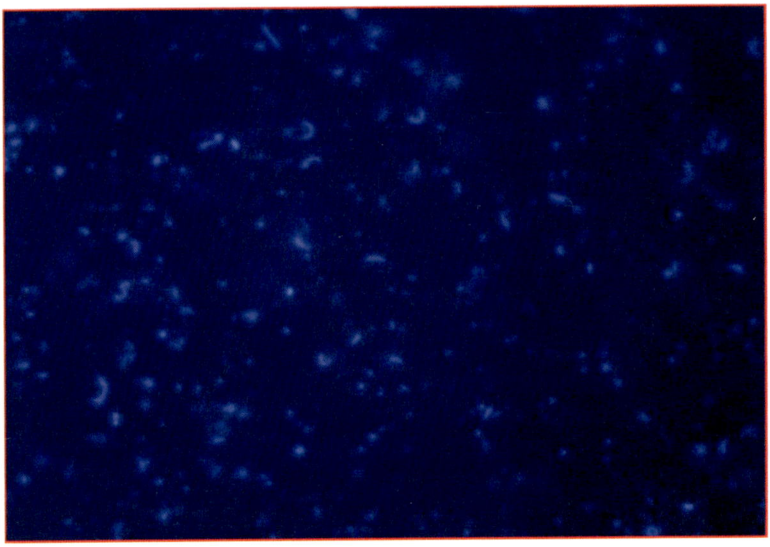

Figure 17.3 .Fluorescence microscopic image of DAPI stained bacterial cells and bacterial cells from drinking water show blue fluorescence on the membrane filter.

PHASE-CONTRAST LIGHT MICROSCOPE

Most of the microorganisms are colorless and also their organelles, which often too difficult to see by dark field microscopy and common bright field microscopy. The observations are clear by aphase-contrast microscope such as the structure of endospores, unstained bacteria and living organisms. The species are often arranged in pairs or chains, rod-shaped and rounded or square ends known as *Bacillus* species and their endospores are mostly oval shape or sometimes cylindrical. *Clostridium* species are repeatedly set in short chains or pairs, mostly with rounded or sometimes pointed ends. The endospores regularly swell up the cell. Consequently, by means of ready slides of *Clostridium* and *Bacillus* the students gain knowledge about a phase-contrast microscope and observing precise bacterial structures such as diverse endospores are present in bacteria. Pond water is generally crowded with protists and bacteria, which observe the internal structure by using phase-contrast microscope. Apart from this we can also see the internal structure of the paramecium (protist) and slowing down the different microorganisms which are present in pond water.

PRINCIPLE

The colorless and transparent living microorganisms having organelles are generally impossible to see by bright-field microscope due to their absorption, refract and reflect limits. Another reason is a sufficient amount of light diffract with the nearby environment for the rest of the microbes. The microbes are visible when they absorb much more light or to refract and reflect from the environment. Thoroughly, we can see microbe organelles by this process. The observations of living and invisible unstained microbes are by using only phase-contrast microscope. The condenser has an annular diaphragm, hollow cone which produces light, glass disk (phase plat) has an objective with a thin film transparent material deposit on it for phase changes produced in the specimen which are accentuated. The phase change of images is observed due to the difference in intensity in light. The positive phase plate is diffracted due to undiffracted light in dark phase-contrast light microscopy or negative phase plate same as above given light relative to the directed light-producing by phase-contrast light microscopy.

REQUIREMENTS

- Pond waste water
- Phase-contrast light microscope
- New slides and coverslips
- Pipette
- Microorganisms
- Methyl cellulose "Protoslo, Carolina Biological Supply"
- Tweezers
- Lens paper

- Lens cleaner
- Prepared slides of "*Bacillus* or *Clostridium*"

PROCEDURE

1. Create waste pond water.
2. Add one drop of methylcellulose to slow down the swimming of the microorganisms.
3. Prepared slides of selected microbes which are given above in the introduction part.
4. The slide is placed on the stage of phase-contrast microscope so that specimen is under the light hole.
5. The lenses are set into place and rotate the position (annular diaphragm) of the specimen if it is required.
6. It is necessary that the cone of light produced by the annular diaphragm below the condenser be centered.
7. Center with the 10X objective and observe the specimen which is under over light.
8. Rotate the annular diaphragm into the correct position for the observation with 40X objective if the image is not clear.
9. Do the same with the oil immersion lens.
10. In the report for exercise 4, sketch several of the microorganisms that you have observed.
11. If you examine pond water (collection of microbes), use the pictorial guides provided by your teacher to help you in identifying microorganisms that are present in pond water.

PRECAUTIONS

- Make sure the specimen is directly over the light hole in the phase-contrast microscope.
- The phase elements must be properly associated.
- Misalignment is the major consequence that beginning students encounter in phase-contrast microscopy.
- If your microscope is not properly worked ask your lab technician for help.
- Keep the microscope covered when not in use

- I should be kept away from moisture and direct sunlight.
- Never clean lenses with a paper towel or cloth.
- Never use a slide without cover glass while examining an object.

VIVA QUESTIONS

1. What is a phase-contrast microscope?
2. What is the principle of this test?
3. Why we use phase-contrast microscope instead of a bright-field microscope?
4. What is the significance of this test?
5. What is paramecium?
6. What is protista? Why we are not used bright-field microscope to see them?
7. What is a microscope?
8. What is the maximum resolution of a microscope?
9. What is the principle of a microscope?
10. What is the meaning of magnification power?
11. What is resolving power?
12. Who discovered the first microscope?

PART-IV
VIROLOGY RELATED PROTOCOLS

CHAPTER 18

GENOMIC DNA ESTIMATION BY AGAROSE GEL ELECTROPHORESIS

INTRODUCTION

Agarose gel electrophoresis (Figure 18.1) is the most effective way of separating DNA fragments of varying sizes ranging from 100 bp to 25 kb.

PRINCIPLE

Agarose is isolated from the seaweed genera Gelidium and Gracilaria and consists of repeated agarobiose (L- and D-galactose) subunits.

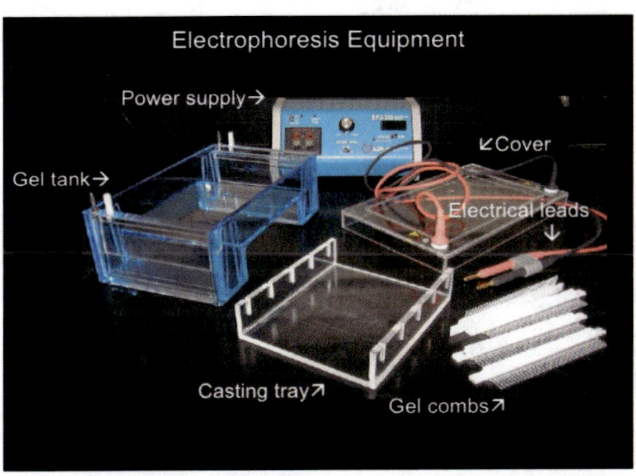

Figure 18.1. Schematic diagram of equipment for gel electrophoresis

REQUIREMENTS

Tris-borate EDTA (TBE electrophoresis buffer) (0.9 M tris HCl, 0.9 M Boric acid, 0.02 M EDTA) 10X :

108 gm of tris base, 55 gm of boric acid and 9.5 gm of EDTA were dissolved in distilled water, the pH 8 was adjusted with 1 N NaOH and HCl and the final volume made to 1 liter. The buffer was autoclaved and kept at room temperature. It was diluted to 0.5 X as according to use.

DNA loading dye:

(0.255 bromophenol blue (w/v), 30% glycerol (v/v), 1% SDS) 0.025 gm bromophenol blue dye and 0.1 gm SDS were dissolved in 3 ml of glycerol and made up to a final volume of 10 ml the container was wrapped in aluminum foil and stored at room temperature.

Ethidium bromide 10 mg/ml:

10 mg of ethidium bromide was dissolved in 1 ml distilled water, stirred vigorously to completely dissolve the dye. The dye was diluted to 1 mg/ml. the container was wrapped in aluminum foil and stored at 4°C.

Agarose 0.8% (w/v): dissolved 0.8 gm agarose in 100 ml 1X TBE buffer.

PROCEDURE

1. Agarose was boiled in 0.5 X TBE buffer in a microwave oven.
2. It was cooled to about 50°C and poured on a gel plate with a comb fixed in gel casting apparatus.
3. The gel was allowed to solidify for about 30 minutes, combs were carefully removed and the gel was kept submerged in an electrophoresis buffer tank.
4. Genomic DNA (10 μl) was mixed with 10 μl of gel loading dye and loaded into agarose gel containing 2 μl of ethidium bromide (1 mg/ml) in 0.5X TBE buffer and run at constant voltage (50 V) for about 2 hours (Figure 18.2).
5. A λ DNA double digest marker was included on one side of the gel as a molecular standard.
6. The gel was visualized on a UV transilluminator (Figure 18.3).

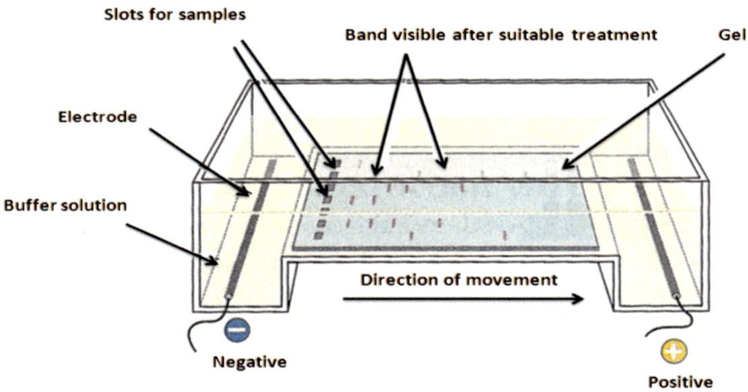

Figure 18.2. Agarose gel electrophoresis of DNA

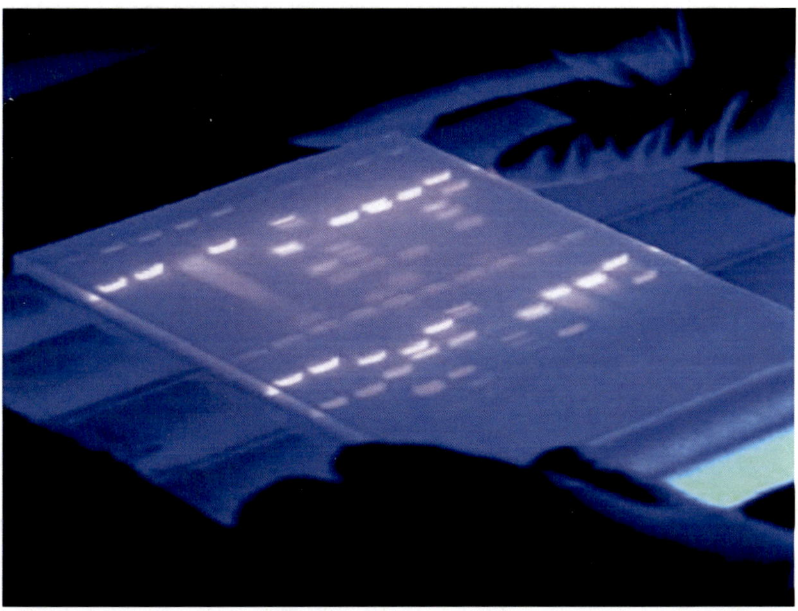

Figure 18.3. Gel picture visualized under gel doc

PRECAUTIONS

- Gel pouring should be very carefully done without entrapping any air bubbles.
- Wear Gloves before the experiment.

- Clean the apparatus only with a cotton cloth.
- Never use alcohol as it cracking the plastic wares.

VIVA QUESTIONS

1. What is agarose gel?
2. How much amount of DNA is separated?
3. What is the principle of this test?
4. Why we use ethidium bromide?
5. Why we perform this test?
6. What is TBE buffer?
7. What is the function of a UV transilluminator?

CHAPTER 19

SODIUM DODECYL SULFATE-POLYACRYLAMIDE GEL ELECTROPHORESIS (SDS-PAGE)

INTRODUCTION

Sodium Dodecyl Sulfate polyacrylamide gel electrophoresis (PAGE) is a commonly used technique for the separation of proteins. This method was first developed by Laemmli in 1970 using the denaturing sodium dodecyl sulfate (SDS) salt. SDS-PAGE technique based on the presence of β-mercaptoethanol and SDS for the denaturation of proteins, the proteins get dissociated into further subunits or monomers and get charged negatively. The uniform charge to mass ratio helps the proteins to flow in an electric field. Proteins are separated according to their mass/size ratio. In SDS-PAGE different compositions with different pH of buffers are used to generate a voltage gradient. The apparatus is used for the SDS-PAGE technique shown in figure 19.1.

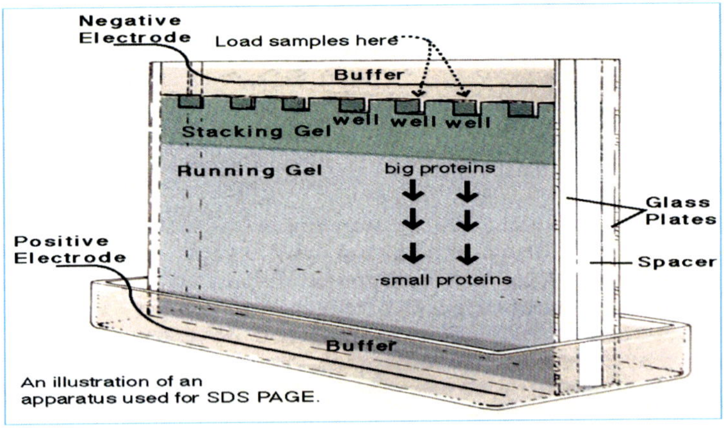

Figure 19.1. SDS-PAGE apparatus.

PRINCIPLE

It is an anionic detergent that strongly binds to the proteins. Approximately the number of SDS molecule bounds to polypeptide chain is half the number of amino acid residues in that polypeptide chain. The complex of protein-SDS carries overall negative charges, hence it moves towards the anode and its separation is totally based on size.

REQUIREMENTS

- 30 % acrylamide-bisacrylamide solution: Dissolves 29 gm of acrylamide and 1 % bisacrylamide in distilled water and make up the volume up to 100 ml.
- 10 % SDS (Sodium Dodecyl Sulfate): Dissolve 10 gm of SDS in 100 ml distilled water.
- 10 % APS (Ammonium Per Sulfate): Dissolve 10 gm APS in 100 ml distilled water (make fresh each time).
- Tetra methyl ethylene diamine (TEMED): Freshly used; Tris-HCl: pH 8.8 (resolving gel): 1 M Tris, pH 6.8 (stacking gel)
- 5x SDS running buffer (1000 ml): Tris 15 gm: 0.05 M; Glycine 30 gm: 0.192 M; SDS 5 gm (pH- 8.4)
- Coomassie briliant blue stain: 10 % (v/v) acetic acid; 0.006 % (w/v) Coomassie Blue dye and 90 % millipore water

- Isopropanol fixing solution: 10 % (v/v) acetic acid; 25 % (v/v) isopropanol; 65 % double distilled H_2O
- SDS sample loading buffer (40 ml): millipore water 16 ml; 0.5 M Tris, pH 6.8 5 ml; 50 % Glycerol 8 ml; 10 % SDS 8 ml; β-mercaptoethanol 2 ml (add immediately before use)
- Sample loading dye: bromophenol blue: 5 mg; 10 % (v/v) acetic acid: 1ml

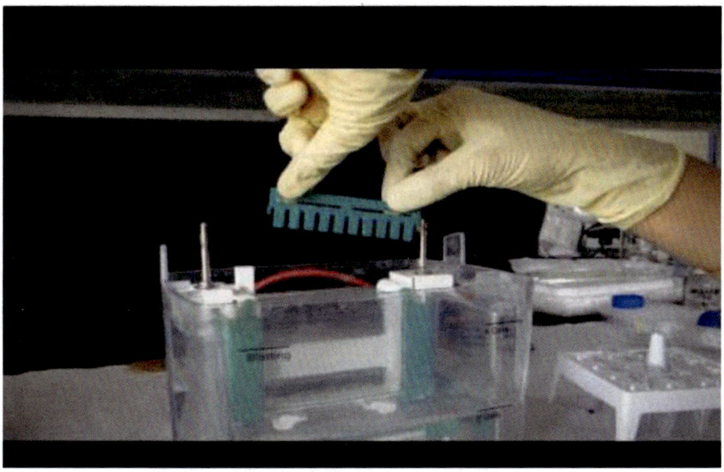

Figure 19.2. Clean glass plates, comb and spacers

PROCEDURE

- Clean and dry the complete unit of SDS (glass plates, comb and spacers) then make them assemble properly before adding sample (Figure 19.2).
- Make 30 ml separating gel (15 %) solution by mixing, acrylamide-bisacrylamide solution (20 ml), tris HCl pH 8.8 (8 ml), millipore water (11.4 ml), APS solution (0.2 ml), SDS (0.4 ml) and TEMED (20 μl).
- Mix carefully and pour this solution between the two glass plates. Pour millipore water (5 ml) above the gel and leave it for 30-60 min.
- Prepare 10 ml stacking gel (4 %) by mixing, acrylamide-bisacrylamide solution (1.35 ml), tris-HCl pH 6.8 (1 ml), millipore water (7.5 ml), APS solution (50 μl), SDS (0.1 ml) and TEMED (10 μl) (Figure 19.3).

- Remove the water layer from the gel and then pour the stacking gel solution.
- After that place the comb in the stacking gel and allows it to solidify approximately (30-60 min) in a cold environment.
- After gel polymerization, remove the comb carefully. Fill the electrophoretic tank with a running buffer and connect the electrodes.
- Prepare protein sample with loading dye and protein molecular weight marker.
- Load them carefully into the wells.
- Turn on the current 30 mA for about 3 hrs until the complete sample reaches the bottom of the glass plates.
- After the three hrs, remove the gel carefully between the glass plates and placed it into a large tray having staining dye solution (500 ml).
- After staining transfer gel into another tray having a destaining solution (250 ml), change this solution several times until the stain is completely removed.
- If it is not removed then dip the gel overnight.
- After de-stained gel now transfers in a poly sheet and scan it with the help of a scanner.
- The scanned gel picture is used for further band quantification (Figure 19.4).

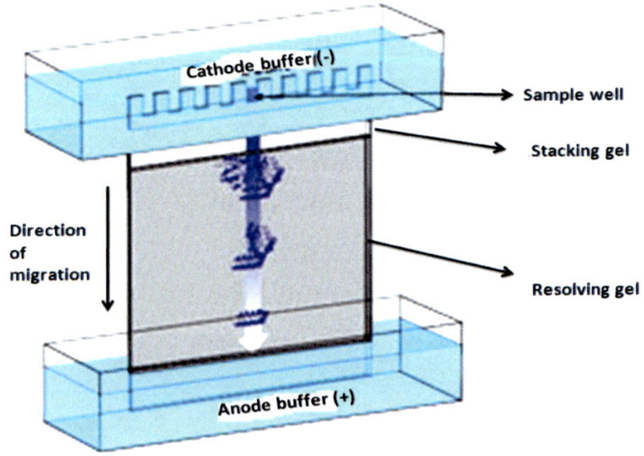

Figure 19.3. Two different gels in SDS-PAGE.

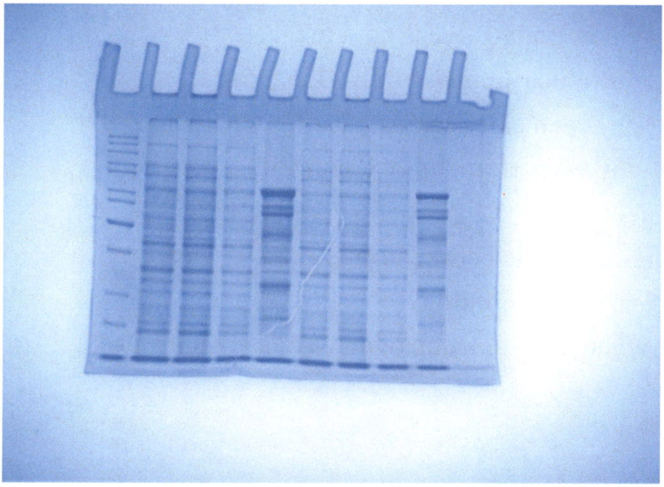

Figure 19.4. Scanned SDS gel picture.

PRECAUTIONS

- Clean and dry the SDS gel apparatus properly before performing this experiment.
- Wear gloves to handle with acrylamide gel.
- Sterile glassware should be used.

VIVA QUESTIONS

1. What is SDS?
2. What is PAGE?
3. What is the meaning of SDS-PAGE?
4. What is protein?
5. Define protein types?
6. What are pH and buffer?
7. Why we perform this test?
8. What is the principle of this test?
9. What is the function of acrylamide bis- solution in SDS-PAGE?
10. What is the role of TEMED in SDS-PAGE?
11. Who discovered SDS-PAGE?
12. What is a running buffer?

13. Which loading dye is used in this test?
14. How much amount of TEMED is used?
15. Why we make stacking gel after separating gel?
16. Why we perform this experiment?
17. What is DNA?
18. Differentiate between DNA and DNA marker?
19. What is TBE buffer?
20. What is the function of a UV transilluminator?

CHAPTER 20

POLYMERASE CHAIN REACTION (PCR)

INTRODUCTION

The Polymerase Chain Reaction (PCR) means DNA amplification from a very less amount of initial DNA concentration in the PCR machine. There are three major steps in PCR reactions it consists:

- Denaturation
- Annealing and
- Extension

1. These steps are repeated approximately 30-40 times, called cycles.
2. In the first cycle, "denaturation" double-stranded DNA gets denatured in high heat about 90°C.
3. The temperature is then cool down up to 40°C to 55°C.
4. This second step "annealing" allows the hybridization of two oligonucleotide primers to bind their complementary sites that flank the target DNA.
5. These annealed oligonucleotide primers provide a free 3′ hydroxyl group for further DNA synthesis with the help of an enzyme DNA polymerase.

6. This further DNA synthesis step is known as the third step “extension”; see Figure 20.1:

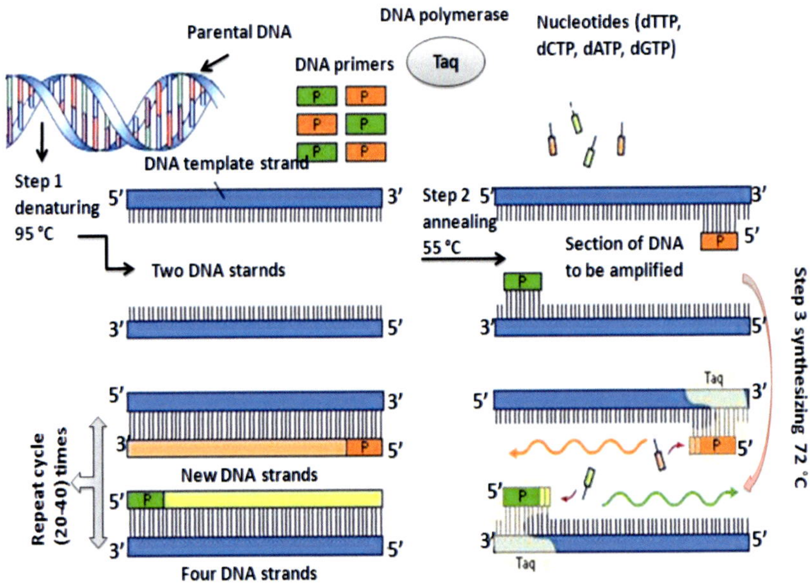

Figure 20.1. Basic stages of PCR reactions

PRINCIPLE

The principle of PCR is based on the amplification of a precise fragment of DNA from a very small amount of starting material. Starting material is usually termed the template DNA and in many cases required DNA purification before amplification. The two oligonucleotide primers may be chemically synthesized, each complementary to the stretch of DNA to the 3’ side of the target DNA, one oligonucleotide for each of the two strands.

REQUIREMENTS

Template DNA: 25 ng

Taq DNA polymerase: The thermostable Taq DNA polymerase with 5′-3′ polymerase with polymerase activity and 5′-3′ exonuclease activity was obtained. The enzyme was supplied in a concentration of 2000 U/ml (2 U-μl^{-1}).

dNTP mix: dNTP mix was prepared to a final concentration of 10 mM., an equal volume of each dNTP solution was mixed to a final concentration of 2.5 mM. It was stored at 20°C.

Taq DNA polymerase buffer (10 X): 100 mM Tris-HCl, 500 mM KCl, 15 mM $MgCl_2$ and 0.1 % gelatin.

Primers: The primers were supplied in the lyophilized condition. The primers were suspended in 100 μl of distilled water. Take 2 μl from stock and add 1 ml of distilled water. The primers were then diluted with distilled water to make the working solution of 25 mM-μl^{-1} concentration.

PCR reaction mixture: A total of 25 μl of the reaction mixture is prepared by mixing the appropriate amount above chemicals.

Table 1. Polymerase chain reaction mixture

S.No.	Components	Actual concentration	Final concen-tration	Volume for 1 reaction
1	DNA template	125 ng	25 ng	05.00 μl
2	dNTP mix	10 mM	2.5 mM	01.00 μl
3	Taq DNA polymerase	03 U/μl	1.5 U/μl	00.50 μl
4	Primers	10 mM	1 mM	01.00 μl
5	Assay buffer	10 X	1X	02.50 μl
6	Sterile distilled water			15.00 μl
			Total =	25.00 μl

PROCEDURE

1. The PCR reactions were performed in a final volume of 25 μl
2. DNA amplification was carried out with the following thermal profilc: initial denaturation at 94°C for 5 minutes followed by 45 cycles of denaturation at 94°C for 1 min annealing at 3°C for 1 min, extension at 72°C for 2 min, and a final extension at 72°C for 5 min.
3. Tm was calculated using the formula [2(A+T) + (G+C)].
4. After completion of the PCR cycle, PCR products were stored at -20°C up to 24 hrs.
5. PCR products were visualized in 1.5 % agarose gel.

PRECAUTIONS

- Gloves should be wearing before the experiment to avoid contamination.
- All the reagents of the PCR should be used on ice.
- The reaction mixture should be prepared just before loading the sample in a gel (freshly prepared).
- Do not store the prepared reaction mixture.
- Use PCR workstation for the preparation of reaction mixture.
- Double distilled water should be used.
- Sterile tips should be used.

VIVA QUESTIONS

1. What is PCR?
2. Explain the full form of PCR?
3. What is Taq DNA polymerase?
4. Define three major steps of PCR?
5. What is a primer?
6. Why we need a primer in PCR?
7. What is the meaning of 5′-3′ exonuclease activity?
8. What is the meaning of lyophilized condition?
9. What is A, T, C, and G?
10. Why we use 1.5% agarose gel for visualized PCR products?
11. What are purines and pyrimidines?
12. What is a nucleotide?
13. What is the difference between DNA and Taq DNA polymerase?
14. Who discovered PCR?
15. How much amount of total volume reactions in PCR?

CHAPTER 21

RNA ISOLATION FROM TRIZOL METHOD (CELLS, TISSUE, VIRUS)

INTRODUCTION

TRIzol Reagent is a ready-to-use reagent, designed to isolate high-quality total RNA (as well as DNA and proteins) from cell and tissue samples of human, animal, plant, yeast, or bacterial origin, within one hour. TRIzol Reagent is a monophasic solution of phenol, guanidine isothiocyanate, and other proprietary components that facilitate the isolation of a variety of RNA species of large or small molecular size. TRIzol Reagent maintains the integrity of the RNA due to the highly effective inhibition of RNase activity while disrupting cells and dissolving cell components during sample homogenization. TRIzol Reagent allows for simultaneous processing of a large number of samples and is an improvement to the single-step RNA isolation method developed by Chomcynski and Sacchi in 1987.

EQUIPMENT AND MATERIALS

- Centrifuge.
- Dry Bath
- Vortex
- Nanodrop machine

- Tips
- Micropipette
- 1.5 ml tubes
- Trizol reagent
- Chloroform
- Isopropanol
- 70% ethanol
- DEPEC water (Fresh)

PRINCIPLE

Total RNA Isolation from cells and tissues for use in PCR analysis. TRIzol reagent is a mono-phasic solution of phenol and guanidine isothiocyanate. During tissue homogenization or lysis, the TRIzol reagent maintains RNA integrity, while disrupting cells and dissolving cell components. The addition of chloroform followed by centrifugation separates the solution into an aqueous phase and an organic phase. RNA remains in the aqueous phase and is recovered by precipitation with isopropanol. The RNA is re-dissolved in Nuclease-free water and store at -80^{0}C for further use.

PROTOCOL

1. Add 0.2 mL of TRIzol Reagent per 1×10^5—10^7 cells directly to the culture dish to lyse the cells. c. Pipete the lysate up and down several times to homogenize.
2. Incubate for 5 minutes to permit complete dissociation of the nucleoproteins complex.
3. Add 0.2 mL of chloroform per 1 mL of TRIzol™ Reagent used for lysis, then securely cap the tube.
4. Incubate for 2–3 minutes.
5. Centrifuge the sample for 15 minutes at 12,000 × g at 4°C. The mixture separates into a lower red phenol-chloroform, and interphase, and a colorless upper aqueous phase.
6. Transfer the aqueous phase containing the RNA to a new tube.
7. Transfer the aqueous phase containing the RNA to a new tube by angling the tube at 45° and pipetting the solution out.
8. Add 0.5 mL of isopropanol to the aqueous phase, per 1 mL of TRIzol™ Reagent used for lysis.

9. Incubate for 10 minutes.
10. Centrifuge for 10 minutes at 12,000 × g at 4°C.
11. Total RNA precipitate forms a white gel-like pellet at the bottom of the tube.
12. Discard the supernatant with a micropipette.
13. Resuspend the pellet in 1 mL of 75% ethanol per 1 ml of TRIzol Reagent used for lysis.
    a. *Note:* a. The RNA can be stored in 75% ethanol for at least 1 year at –20°C, or at least 1 week at 4°C.
    b. Vortex the sample briefly, then centrifuge for 5 minutes at 7500 × g at 4°C.
    c. Discard the supernatant with a micropipette.
    d. Vacuum or air dry the RNA pellet for 5–10 minutes.
    e. *Note:* Do not dry the pellet by vacuum centrifuge. Do not let the RNA pellet dry, to ensure total solubilization of the RNA. Partially dissolved RNA samples have an A230/280 ratio <1.6.
14. Resuspend the pellet in 20–50 μL of RNase-free water.
15. Incubate in a water bath or heat block set at 55–60°C for 10–15 minutes. Proceed to downstream applications, or store the RNA at –70°C.
16. Calculate the RNA concentration using the formula A260 × dilution × 40 = μg RNA/mL. Calculate the A260/A280 ratio.
17. A ratio of ~2 is considered pure.
18. RNA samples can be quantified by absorbance without prior dilution using the NanoDrop Spectrophotometer. Refer to the instrument's instructions for more information.

REFERENCES

1. Chomczynski, P., and Sacchi, N. 1987 Single Step Method of RNA Isolation by Acid Guanidinium Thiocyanate-Phenol-Chloroform Extraction. Anal. Biochem. 162, 156-159.

2. Hummon, A. B., Lim S. R., Difilippantonio, M. J., and Ried, T. 2007 Isolation and solubilization of proteins after TRIzol® extraction of RNA and DNA from patient material following prolonged storage. BioTechniques 42, 467-472.

CHAPTER 22

ONE STEP RT-PCR

INTRODUCTION

The Real-Time RT-PCR Assay is used in rRT-PCR on Real-Time PCR Instrument. The Real-Time RT-PCR Assay includes a set of oligonucleotide primers and dual-labeled hydrolysis (Taqman) probes for *in vitro* qualitative detection of virus from serum or plasma collected from human patients with signs and symptoms consistent with dengue (mild or severe).

EQUIPMENT AND MATERIALS

- Spin
- Vortex
- Nanodrop machine
- RT-PCR reagents for (Master mix)
- Tips
- Micropipette
- 1.5 ml tubes
- DEPEC water (Fresh)
- Real-time PCR Machine

PRINCIPLE

Real-Time PCR has become an increasingly popular technique for the analysis of gene expression.

Master mix Per reaction

Components	**Volume (µl)**
4X TaqmanRT-PCR mix	5
40X Taqman RT Enzyme mix	0.25
denUTR-F	0.4
denUTR-R	0.4
denUTR-probe (FAM)	0.4
Primer probe mix GAPDH (HEX)	0.5
Template RNA (100 ng)	4
DEPEC-H2O	9.05
Total	**20 µl**

RT-PCR Program

Temperature (^{0}C)	**Time**	**Cycle**
48	15 min	1
95	10 min	1
95	15 sec	40
60	1 min	

PROTOCOL

1. Isolate the RNA from different dengue infected samples and take one sample as a positive which is uninfected (control).
2. Quantify the RNA concentration by Nanodrop.
3. Calculate the RNA concentration according to 100 ng of RNA which is used for RT-PCR.
4. While doing the RT-PCR always take a negative and positive control assay.
5. Depending on the number of samples make the master mix according to the reaction and add the template in respected wells (Total volume is 20 µl).

6. Prepare RT-PCR plate (96 well) or RT-PCR tubes with master mix and different RNA samples. The total reaction volume is 20 μl according to the experiment.
7. After this run the RT-PCR assay according to the program.
8. Analyze the data with different Ct values.

HOW TO OPERATE BIORAD MACHINE

i. "Switch On" Biorad machine and Laptop with properly engaged charger otherwise you lost your data if Laptop not charged or stop passing electricity.
ii. Click On "BIORED CFX Manager RT-Machine software" on a Laptop.
iii. You will get an RT-PCR software window.
iv. After this, click on User-defined so you will get a small open window.
v. Click on "Select existing" (at the left side) if program set (go to D-Drive), go to Dr. Guru Lab folder (e.g Khushboo Chaudhary folder), get Taqman RNA CT kit (If you follow kit) or Create new if you want to run the new program.
vi. Now click On the "Next" button and then Click on create new (Select the column-it depends on how many samples you want to run), Select on the "Right side" button (Unknown-if you write down about the samples in your data book otherwise naming about all the samples), Now select Fluorophore (HEX-GAPDH probe and FAM-Dengue probe), click on "OK". Strictly follow this otherwise you will not get proper results (especially in the case of dengue). Otherwise, you select other probe box (it depends on which probe is used by you) so remember carefully this step.
vii. Select Target Name e.g Actin or GAPDH.
viii. Click on "HEX" on the left side.
ix. Now, Click on the "OK" Button, this go to "Yes" (Save the plate program where you make your folder).
x. Click On "Open the lid" at the left side button and set the PCR tubes according to the select plate number and after the tube set up in the box machine "Close the Lid".
xi. Now you will found the "NEXT" button and then "Run" (you will get save the program window so you saved "PCRD" file (Need this

file for analysis of data) otherwise you lose your data (This is the main step you follow carefully).

xii. After saving the file, you get the initializing symbol so you noted the time when your program will complete.

xiii. If you want to see it then you go to "Run program" "Time" above the side of the window.

IMPORTANT POINTS

One-step Real-time qRT-PCR Assays

- One-step real-time qRT-PCR reactions were performed one-step RT-PCR Kit.
- Total RNA extracted from the sample was used.
- Real-time qRT-PCR cycling was performed on the Bio-Rad system as follows
- The fluorogenic signal emitted was collected at the end of the annealing-extension step.
- A threshold was automatically set and the threshold cycle value (C_t) was consequently determined.
- Three replicates of the assay within or between runs were performed.

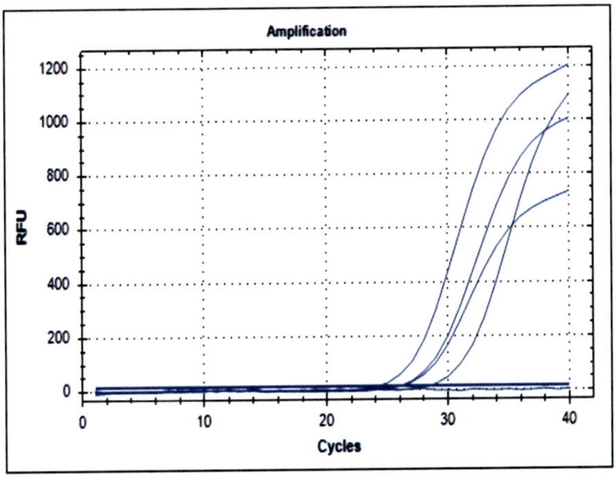

REFERENCES

1. Bustin S. (ed) (2004) A-Z of Quantitative PCR. IUL Biotechnology Series, International University Line, La Jolla, California.

CHAPTER 23

MTT ASSAY

INTRODUCTION

The MTT assay has been widely used to assess cell viability. However, one must consider that the enzymatic reduction of 3-[4,5-dimethylthiazole-2-yl]-2,5-diphenyltetrazolium bromide (MTT) to MTT-formazan is catalyzed by mitochondrial succinate dehydrogenase. Hence, the MTT assay is dependent on mitochondrial respiration and indirectly serves to assess the cellular energy capacity of a cell. The MTT assay is a colorimetric reaction that can easily be measured from cell monolayers that have been plated in 35 mm dishes or multiwell plates.

EQUIPMENT AND MATERIALS

- 96 multiwell plates for cell culture
- Multichannel pipette or automated pipetting station for reagent delivery
- Device (plate shaker) for mixing multiwell plates
- Spectrophotometer, or CCD camera or imaging device capable of reading multiwell plates

MTT REAGENT

- Add 250 mg MTT (Sigma) in 50 ml of distilled water. Filter sterile using 0.22 μm syringe filter

PRINCIPLE

- This colorimetric assay is based on the reduction of a yellow tetrazolium salt (3-(4,5-dimethylthiazol-2-yl)-2,5-diphenyltetrazolium bromide or MTT) to purple formazan crystals by metabolically active cells. The viable cells contain NADPH-dependent oxidoreductase enzymes which reduce the MTT to formazan.

PROTOCOL

- Make plate 20,000 thousand cells per well in a 96 well plate and incubate with appropriate stimulus for the desired time for 48 hours (Depends on cells 90% confluent cells.)
- Cells in 96-well plates were treated with 2-fold dilutions of Drug (according to the experiment) or any compound or dimethyl sulfoxide (DMSO) control, in triplicates, in a total of 100 μl growth medium.
  - *Note:* In the first row 180 μl serum-free media +20μl drug in the first row and 100 μl media in a remaining row or well so the total amount is 200 μl in the first row at 1:10 dilution.
  - *Note:* Maintain 100 μl media in each well and make 2 fold dilutions from 1:10 dilutions so transfer 100 ul in each well to maintain 2 fold dilutions.
- Put this 96 well plate in CO_2 incubation for 4 days.
- After 4 days, An amount of 20 μl of freshly made 5 mg/ml MTT [3-(4,5-dimethyl-2-thiazolyl)-2,5-diphenyl-2H-tetrazolium bromide] solution was added to each well, and cells were incubated at 37°C for 5 h.
- After removal of the medium carefully (without damage monolayer of cells).
  - *Note:* Use multichannel micropipette for removing media.

- Add 200 μl of DMSO to each well to dissolve the purple formazan crystals product and the plates were incubated at 37°C for another 5 min to remove any air bubbles.
- MTT signals were measured photometrically at an absorbance of 550 nm.
- The 50% cytotoxic concentration (CC_{50}) was calculated as the concentration needed to reduce cellular viability to 50%.

INTERPRETING RESULTS

The absorbance reading of the blank must be subtracted from all samples. Absorbance readings from test samples must then be divided by those of the control and multiplied by 100 to give percentage cell viability or proliferation (see formula below). Absorbance values greater than the control indicate cell proliferation, while lower values suggest cell death or inhibition of proliferation.

REFERENCES

1. Mosmann, T. et al. (1983) J. Immunol. Methods 65, 55–63.
2. Tada, H. et al. (1986) J. Immunol. Methods 93, 157–165.
3. Denizot,F. & Lang, R. (1986) J. Immunol. Methods 89, 271–277.
4. Gerlier, D. & Thomasset, N. (1986) J. Immunol. Methods 94, 57–63.
5. Hansen, M. B., Nielsen, S. E. & Berg, K. (1989) J. Immunol. Methods 119, 203–210.

CHAPTER 24

ANTIVIRAL ACTIVITY

INTRODUCTION

Antiviral is an agent that kills a virus or that suppresses its ability to replicate, hence, inhibits its capability to multiply and reproduce. It acts by inhibiting the multiplication of the virus. It is used to lessen the severity of the disease, particularly in individuals at high-risk such as those who are immunosuppressed or in a nursing home. A virus is a just genetic material, DNA or RNA, perhaps with a few enzymes, wrapped in a protein coat. A viral is technically not alive which makes it hard to kill. Further, viruses replicate (make copies of themselves) by hijacking the machinery of the cell they infect, so it is difficult to kill the virus without killing the cell. Antiviral drugs do not destroy their target pathogen; instead, they inhibit their development. Antiviral drugs are a class of medication used for treating viral infections. Most antivirals target specific viruses, while a broad-spectrum antiviral is effective against a wide range of viruses.

EQUIPMENT AND MATERIALS

- Inverted Microscope
- PBS at Room temperature.
- 1.5 ml tube
- Vortex

- Micropipette
- Tips
- Tube stand
- Rocker
- 90% confluent cells
- Culture Medium depends on the cell type
- FBS
- CO_2 incubator

PRINCIPLE

A molecule can act as an anti-viral drug if it inhibits some stage of the virus replication cycle, without being too toxic to the body's cells. . Antiviral compound would include such parameters such as inactivate the extracellular virus particles. Prevent viral attachment to host cell and/or entry, Prevent replication of the viral genome in the host cell, Prevent synthesis of specific viral protein(s)., Prevent assembly or release of new infectious virions in the cytoplasm of the host cell.

PROTOCOL

1. Make 48 well plate with 40, 000 cells in each well and keep plate in CO_2 incubator for one day or it depends on 90% confluent cells.
2. The next day remove the media carefully.
3. Wash the cells twice with 300 μl PBS.
4. Add virus in each well at a particular MOI (It depends on the experiment) and incubate for 1 hr on a rocker. (For 48 well plate volume should be 100 μl)
5. After 1 hr, wash the plate with PBS at least 3 times.
6. An appropriate amount of drug is added to serum-free media in 1.5 ml tubes.
7. Add drug and media 300 μl in each well and it should be in triplicate form.
8. For control add an equal volume of drug diluents such as DMSO in 2 % Medium.

9. Put this plate in a CO_2 incubator for 24 hr incubation.
10. Take supernatant after 24 hpi and do plaque assay for quantification of virus titer.

REFERENCE

1. Morales D. J., Lenschow D. J. (2013). The antiviral activities of ISG15. *J. Mol. Biol.* 425 4995–5008. 10.1016/j.jmb.2013.09.041

CHAPTER 25

CELL VIABILITY ASSAY

INTRODUCTION

To measure cell viability, and cytotoxicity are commonly used to monitor the response and health of cells in culture after treatment with various stimuli. The proper choice of an assay method depends on the number and type of cells used as well as the expected outcome Cell counting using viability dyes such as trypan blue or Calcein-AM can provide both the rate of proliferation as well as the percentage of viable cells.

EQUIPMENT AND MATERIALS

- Opaque-walled multiwell plates adequate for cell culture
- Multichannel pipette or automated pipetting station for reagent delivery
- Device (plate shaker) for mixing multiwell plates
- Luminometer, CCD camera or imaging device capable of reading multiwell plates
- Optional: ATP for use in generating a standard curve)

REAGENT PREPARATION

- Thaw the CellTiter-Glo® Buffer, and equilibrate to room temperature prior to use. For convenience, the CellTiter-Glo® Buffer may be thawed and stored at room temperature for up to 48 hours prior to use.
- Equilibrate the lyophilized CellTiter-Glo® Substrate to room temperature prior to use.
- Transfer the appropriate volume (10ml for Cat.# G7570 and G7571, or 100ml for Cat.# G7572 and G7573) of CellTiter-Glo® Buffer into the amber bottle containing CellTiter-Glo® Substrate to reconstitute the lyophilized enzyme/substrate mixture. This forms the CellTiter-Glo® Reagent.
- Mix by gentle vortexing, swirling or inverting the contents to obtain a homogeneous solution. The CellTiter-Glo® Substrate should go into solution easily in less than 1 minute.

PRINCIPLE

Cell-based assays are often used for screening collections of compounds to determine the test molecules have effects on cell proliferation or show direct cytotoxic effects that lead to cell death. There are a variety of assay methods that can be used to estimate the number of viable eukaryotic cells. It will provide some of the major methods used in multi-well formats where data are recorded using a plate reader. They are based on various cell functions such as enzyme activity, cell membrane permeability, cell adherence, ATP production, co-enzyme production, and nucleotide uptake activity.

PROTOCOL

- Make plate using 20,000 A549, Huh-7 and vero cells per well in a 96-well plate.
- Multiwell plates must be compatible with the luminometer used.
- Prepare control wells containing medium without cells to obtain a value for background luminescence.
- Prepare two-fold serial dilutions of the test compound. To determine CC50 five different concentrations of test compounds should be used.

- Add the test compound to experimental wells, and incubate according to culture protocol (Incubate at least for 24 hr)
- Equilibrate the plate and its contents at room temperature for approximately 30 minutes.
- Add a volume of CellTiter-Glo® Reagent equal to the volume of cell culture medium present in each well (e.g., add 100µl of the reagent to 100µl of medium containing cells for a 96-well plate, or add 25µl of the reagent to 25µl of medium containing cells for a 384-well plate).
- Mix contents for 2 minutes on an orbital shaker to induce cell lysis.
- Allow the plate to incubate at room temperature for 10 minutes to stabilize the luminescent signal.

Note: Uneven luminescent signal within standard plates can be caused by temperature gradients, uneven seeding of cells or edge effects in multiwell plates. Record luminescence.

Note: Instrument settings depend on the manufacturer. An integration time of 0.25–1 second per well should serve as a guideline.

PROTOCOL FOR GENERATING AN ATP STANDARD CURVE (OPTIONAL)

It is a good practice to generate a standard curve using the same plate on which samples are assayed. We recommend ATP disodium salt (Cat.# P1132, Sigma Cat.# A7699 or GE Healthcare Cat.# 27-1006). The ATP standard curve should be generated immediately prior to adding the CellTiter-Glo® Reagent because endogenous ATPase enzymes found in sera may reduce ATP levels.

1. Prepare 1µM ATP in culture medium (100µl of 1µM ATP solution contains 10–10 moles ATP).
2. Prepare serial tenfold dilutions of ATP in culture medium (1µM to 10nM; 100µl contains 10–10 to 10–12 moles of ATP).
3. Prepare a multiwell plate with varying concentrations of ATP standard in 100µl medium (25µl for a 384-well plate).
4. Add a volume of CellTiter-Glo® Reagent equal to the volume of ATP standard present in each well.

5. Mix contents for 2 minutes on an orbital shaker.
6. Allow the plate to incubate at room temperature for 10 minutes to stabilize the luminescent signal.
7. Record the luminescence.

CHAPTER 26

VIRUS BUDDING ASSAY

INTRODUCTION

Equipment and Materials

- Vortex mixer
- Pipettes
- Autoclaved Tips
- Deionized water
- CO_2 incubator
- Micropipette
- PBS
- 90% confluent cells
- Virus

PRINCIPLE

The release can occur in several different manners, depending upon the virus. Viruses that obtain their envelope from the plasma membrane generally assemble on the inside layer of the plasma membrane, embedding their envelope proteins into the plasma membrane. As the viral capsid proteins interact, the membrane-associated viral proteins cause the plasma membrane to begin curving around the capsid. This continues until the plasma membrane is completely wrapped around the virus, which leaves the cell.

PROTOCOL

1. Seed the cells according to your plan for the experiment (90% confluent cells used).
2. Keep the plate out and check cells before starting your experiment.
3. Remove media and infect the cells with M.O.I (experiment plan) and incubate at 37^0C for 1 hr (attachment).
4. After 1 hr completed, keep out the plate from the CO_2 incubator.
5. Wash the plate with PBS at least 5 times and add serum-free media and incubate for 2-4 hr in a CO_2 incubator.
6. Keep out the plate from the CO_2 incubator after incubation time is completed and wash with cold PBS.
7. Add serum-free media or add (serum-free media+drug= according to experiment).
8. Keep this plate in a CO_2 incubator for 30 min.
9. After the incubation time period, keep out the plate from the incubator.
10. Take soup or supernatant at 30 min,1 hr, 2 hr, 4 hr (depends on experiment and virus life cycle).
11. Virus released at 1 and 2 h (PPRV) or 30 min is quantified by plaque assay.

REFERENCES

1. Riss TL, Moravec RA. Use of multiple assay endpoints to investigate the effects of incubation time, dose of toxin, and plating density in cell-based cytotoxicity assays. Assay Drug Dev Technol. 2004;2(1):51–62. [PubMed]
2. Duellman SJ, Zhou W, Meisenheimer P, Vidugiris G, Cali JJ, Gautam P, Wennerberg K, Vidugiriene J. Bioluminescent, Nonlytic, Real-Time Cell Viability Assay and Use in Inhibitor Screening. Assay Drug Dev Tech. 2015;13(8):456–65. [PMC free article] [PubMed]

CHAPTER 27

BIOSAFETY GUIDELINES INTELLECTUAL PROPERTY RIGHTS AND ENTREPRENEURSHIP DEVELOPMENT

With the growing concerns arising from genetically modified organisms throughout the world the WHO/UNIDO/FAO/UNEP has built up an informal working group on Biosafety. The group prepared in 1991 the "Voluntary Code of Conduct for the Release of Organisms into the Environment". The ICGEB has also played important role in issues related to biosafety and the environmentally sustainable use of biotechnology. The organization is ICGEB has annual workshops on biosafety and on risk assessment for the release of GMOs. It collaborates with the management of UNIDO's BINAS (Biosafety Information Network and Advisory Service)., aimed at monitoring the global development in regulatory issues in biotechnology. The ICGEB has also provided an online bibliographic database on biosafety and risk assessment for the environmental release of GMOs.

TOPIC OF CONCERNS

General Concern

- Higher cost of agriculture
- Ethical Issues
- Loss of familiarity

Risks for the Environment

- Unpredictable gene expression
- Instability of transgene
- Increased use of chemicals in agriculture
- Resistant of targeted organisms
- Susceptibility of non-target organisms

Risks for Human Health

- Pathogens
- Drug Resistance
- Antibiotic Resistance
- Allergies
- Food Quality
- Food Safety
- Toxicity

Risks of Interaction with Non-target Organisms

- Generation of New Line viruses by recombination
- Horizontal Gene transfer
- Genetic Pollution through pollen or seed dispersal
- Transfer of foreign gene to microorganisms i.e. DNA uptake

Biosafety Guidelines and Regulation

There are many countries that have formulated the biosafety guidelines for rDNA manipulation with the aims: Minimize the probability of occasional release of GMMs. Ban the deliberate release of such organisms into the environment. In India, DBT has evolved "the recombinant DNA safety guidelines to exercise powers conferred through the environmental protection ACT 1986 for the manufacture, use, import, export and storage of hazards microorganisms/ genetically engineered organisms, cell ,etc. The guidelines are being implemented through the following three mechanisms such as:

The genetic engineering approval committee (GEAC) of the Ministry of Environment and Forest has the power to permit large-scale use of GMOs at the commercial level, and open field trials of transgenic materials including agricultural crops, industrial products, health care ,etc. (Annual report 1995-96 DBT).

The institutional biosafety committee (IBSCs) monitors the research activity at the institutional level.

The review committee on genetic manipulation (RCGM) functioning in the DBT allows the risky research activities in the laboratories.

Main Function of Institutional biosafety committees (IBSC)

- To note
- To approve
- To see the approval of RCGM and to recommend

Main Function of the Review committee on genetic manipulation (RCGM)

- To recommend the generation of appropriate biosafety and agronomic data
- To approve
- To note

Main Function of the Genetic engineering approval committee (GEAC)

- Inform decision to administrative ministry to follow PVP/Seed Act
- To approve open release to the environment
- To approve for large scale use.

Main Function of the Indian council of Agriculture Research (ICAR)

To generate complete agronomic data on transgenics

To recommend suitable transgenics for commercial release

Main Function of the Monitoring and Evaluation Committee (MEC)

Recommend safe and agronomically viable transgenics

Inspect facility

Analyze data

Visit trial sites

Intellectual Property Rights

The physical objects such as household goods or land are the properties of a person. Similarly, a country has its own property. The ownership and the rights on the property of a person are protected by certain laws operating in the country. This type of physical property is tangible. On the other hand, the transformed microorganism, plants , animals and technologies for the production of commercial products are exclusively the property of the intellectuals.

The rights of the intellectual must be protected and it does by certain laws framed by a country. However, it is important to differentiate between physical property and intellectual property. For example, the seed of a plant is a tangible asset; it can be sold in the market and money can be made from it. But the intellectual property is an intangible asset.

Legal rights or patents provide an inventor only a temporary monopoly on the use of an invention, in return for disclosing the knowledge to the others in a specification that is intended to be both comprehensive to, and experimentally reproducible by a person skilled in the art. The laws are formulated from time to time at national and international levels. The USA has declared for adopting strong and uniform IPR laws thought out the world.

Development of the crop variety is another intellectual property right. It is protected by "Plant Breeders rights" (PBRs). The PBRs are available in developed countries but not in India. The plant breeders rights recognize the impact that farmers and rural communities have contributed lot to the creation, conservation, exchange and knowledge of genetic/ species utilization of genetic diversity. The intellectual property rights and IPP granted by the government to plant breeders are to exclude others for about 15-20 years from producing or commercializing materials of a specific plant variety. But this variety should be now, novel and never existing before.

In this biotechnology has played a significant role in providing processing, designing and production of valuable commercial products utilizable in many areas of the society as well as the country such as medical, agriculture, health care, industry, environment ,etc. The technology transfer in biotechnology requires a minimum amount of technical and legal capability which the developing countries lack at present.

Stages of Protection

Intellectual property rights are protected in different ways such as patents, copyrights, trade secrets and trademarks, designs, geographical indications.

Patents

It is a special right to the inventor that has been granted by the government through legislation for trading new articles. A patent is a personal property that can be licensed or sold by the person/organization just like any other property. For example, Alexender Grahm Bell obtained a patent for his telephone. This gave him the power to prevent an engine from being made or using or selling a telephone elsewhere. In the USA the maximum limit of this monopoly is for 17 years. In India the Indian patent Act passed in 1970 for the "process of patents" but not the "product patent" and the maximum duration of a patent is 5 years from the date of grant and 7 years from the date of filing the patent application. The least duration between five years to seven years is applicable for patents. The patents in terms give the inventor the right to exclude the others from making, using or selling his invention as disclosed in "claims" of the patent. Obviously, it is difficult to keep secret certain inventions such as the fermentation process. Therefore guidance should be obtained from a qualified patent attorney.

Copyrights

This is protected by only the expression of ideas. One of the best examples of copyrights is books. The authors, editors, publishers or both the publisher and author/editor have copyrights. The materials of the books cannot be reprinted or reproduced without written permission from copyright holders. The copyrights protected the expressed materials like materials in printed, recorded video and taped forms. Biotechnological materials subject to copyrights include a database of DNA sequences or any published forms. In India, the copyrights act passed or was amended in 1994 and brought enforce in 1999. It includes computer programs, tables and databases. The computer program is defined as a set of works expressed in words, schemes, codes and any other form of including a machine-readable medium capable of causing a computer to perform a particular task or achieve a particular result. The

Ministry of Human Resource and Development look after copyright act in India.

Trademarks

It is an identification symbol that is used in the course of trade to enable the public to distinguish trader's goods from the similar goods of the other traders. The public makes use of these trade works in order to choose whose goods they will have to buy. If they are satisfied with the purchase, they can simply repeat their order by using the trademark, for example, KODAK for photography goods, IBM for computers, Zodiac for readymade clothes, etc. The laws vary in every country. Through agreement, it is ensured that the trademark of one country must be protected in another country. Therefore, there are a number of multinational companies spend a large amount of money to maintain their trademarks throughout the world.

World Intellectual Property Organisation (WIPO)

It is one of the specialized agencies of the United Nations. It has provided that the intellectual property shall include rights relating to the following: artistic, literary and scientific works, the performance of artists, phonograms, broadcast; innovation in all field of human endeavor; scientific discoveries; trademarks; service marks and commercial names; industrial designs; protection against unfair competition and all other rights resulting from intellectual activity in the area of industrial scientific literary or artistic fields. Intellectual property rights are protected and governed by appropriate national legislation.

General Agreement of Tariffs and Trade (GATT) and Trade Related IPRs (TRIPs)

It was framed in 1948 by developed countries to settle the disputes among the countries regarding the share of world trade. It is decided by tariff rates and quantitative restrictions on imports and exports. For a long time benefits from GATT were achieved only by developed countries. In 1988, the US Congress enacted a law " the Omnibus Trade and Competitiveness Act (OTCA). The USA becomes powerful to investigate the laws related to trade and check them within the desired period, the US takes action against that country. In 1992, the US gave a warning to India to change some of its laws of IPR, patents and copyrights. A certain inhibition was acted in India to sign on GATT draft. Therefore, There was much debate throughout the country on this issue and the bad intention of the US. The scientist, politicians and professionals argued that the total package of TRIPS must guarantee for economic and technological subjugation of the country. There are many groups and organizations that have rejected this draft of suggestion and opposed the decision taken by the government.

Index

R

S

T

U

V

W

Y

Z

Printed in the United States
by Baker & Taylor Publisher Services